# Federal Reinsurance for Disasters

September 2002



The Congress of the United States ■ Congressional Budget Office

# Note

The photo on the cover is from the *San Francisco Chronicle.* It shows vehicles trapped on an elevated section of Interstate 5 north of Sylmar, California, following the Northridge earthquake of January 17, 1994.

# Preface

This Congressional Budget Office (CBO) study, which was prepared at the request of the Senate Budget Committee, analyzes proposals for federal reinsurance of risks from terrorism and natural disasters. David Torregrosa of CBO's Microeconomic and Financial Studies Division wrote the paper, with contributions from Perry Beider, Kim Kowalewski, and Marvin Phaup, under the direction of Roger Hitchner. Tim VandenBerg, formerly of CBO, began the project.

Many other people contributed to the report. Rade T. Musulin of the Florida Farm Bureau Insurance Companies; Richard Roth Jr., a consulting actuary; Howard Kunreuther of the University of Pennsylvania's Wharton School; and Ron Feldman of the Minneapolis Federal Reserve reviewed several drafts. David Rafferty, formerly of the New Zealand Treasury, provided valuable information on New Zealand's disaster insurance program. Barry Anderson, Megan Carroll, Deborah Clay-Mendez, William Gainer, Mark Hadley, Arlene Holen, Deborah Lucas, Angelo Mascaro, Robert Murphy, Allison Percy, and Tom Woodward, all of CBO, made useful suggestions.

Amelie Cagle, Mark Puccia, Donald Watson, and David Weiss, all of Standard & Poor's, provided background information on the insurance industry and the implications of September 11. Anthony Mormino, Frank Nutter, and Mary Zetwick of the Reinsurance Association of America provided information on the reinsurance industry. Jim Boyd of Resources for the Future provided guidance on the effects of limiting damage awards.

The following people also contributed various forms of assistance or data: Dennis Kuzak of EQECAT; Barbara Jacobik, Tim Mantz, and Rick Shiver of the Federal Emergency Management Agency; Herbert Mitchell and Joyce Northwood of the Small Business Administration; Dan Kutzer of the Florida Windstorm Underwriting Association; J.D. Lester of the Florida Residential Property and Casualty Joint Underwriting Association; Jack Nicholson of the Florida Hurricane Catastrophe Fund; Stan Devereux of the California Earthquake Authority; Richard Kerr and Michelle Kinlaw of MarketScout; Barry Meiners of the Council of Insurance Agents and Brokers; and John Rollins of the Florida Farm Bureau Insurance Companies.

Christine Bogusz edited the report, and Leah Mazade proofread it. Rae Wiseman produced drafts of the manuscript, and Kathryn Winstead prepared the study for publication. Annette Kalicki produced the electronic versions for CBO's Web site.

Dan L. Crippen
Director

September 2002

This study and other CBO publications
are available at CBO's Web site:
www.cbo.gov

CONTENTS

## Tables



## Figure



## Box

# Summary and Introduction

The terrorist attacks of September 11, 2001, inflicted enormous personal and property losses on the United States. Insurance payments for economic losses on that day will surpass $30 billion and could top $50 billion. Among the many adverse economic effects that resulted from the attacks is a continuing shortage of insurance against losses of property from terrorism. Coverage is difficult to obtain, especially for buildings or sites that are landmarks; where coverage is available, prices are high and restrictions are numerous.

Policymakers have developed two types of federal proposals to increase the supply of property and casualty insurance. Under one set of options, which was created following Hurricane Andrew in August 1992 and the Northridge earthquake in January 1994, the federal government would auction reinsurance contracts to primary insurance companies and state-sponsored insurers. Reinsurance is an established practice among private insurers. By buying reinsurance, primary insurers spread the risk of loss more widely and strengthen their ability to insure against catastrophes. The intent of those proposals is to offer federal reinsurance when coverage is in short supply, at market prices that are expected to cover the government's costs. By contrast, under proposals developed since September 11, the government would pay for most losses from a terrorist attack directly, without reimbursement or with only partial reimbursement. Even though one type of proposal was created in response to natural disasters and the other from an act of terrorism, both types could be considered viable alternatives, whatever the source of catastrophic loss.

A key consideration in Congressional deliberations about disaster reinsurance is how the property and casualty industry would respond without federal intervention. That is, would the private supply of insurance rebound quickly? A fast recovery would counter a potential slowdown in construction, an industry in which insurance is often required for financing. Alternatively, is the risk of terrorism uninsurable, such that supply could not recover without the government's assistance? Furthermore, given that the Congress may need to act without fully understanding the industry's ability to respond, are there policies that could avoid undermining private activity while providing a backstop to private efforts if they proved inadequate? How much would such policies cost taxpayers and the government?

This Congressional Budget Office (CBO) study examines the market for property and casualty insurance; the market's response to recent large, unanticipated losses; and policies that the Congress is considering to increase the availability of insurance. In brief:

- Large insured losses that are unanticipated reduce the supply of insurance and put pressure on private insurers to raise their prices, primarily by reducing insurers' net worth and by increasing their uncertainty about future losses.

- Higher prices for insurance eventually attract additional investment into the disrupted market. As a result, insurers reassess their risks, supply increases, and prices decline to levels that are consistent with those revised perceptions of risk. But that adjustment takes time, during which some property owners are unable to get the insurance they want at prices they are willing to pay.

- Proponents of a federal reinsurance program contend that the government needs to add capital to the market quickly during times of disruption and to withdraw it as the market recovers. Those actions would moderate fluctuations in the availability of insurance, address the possibility that private insurance might prove incapable of surviving the "Big One" (usually defined as a hurricane or earthquake costing in the $100 billion range), and accomplish both without displacing private supply. Federal reinsurance might also reduce the need for federal assistance after a disaster.

- A federal program to expand the supply of property and casualty insurance would probably expose taxpayers to substantial risk—up to $100 billion in contingent liabilities for terrorism insurance and up to $25 billion for natural disaster risk—and raise expected budget outlays. In addition, subsidies could lead to fewer preventive actions by insured parties and also to more risk taking, which could boost losses from a disaster. A federal program could also delay innovation by private insurers and the capital markets that could increase private supply.

- States' experience with regulations has shown that controlling prices and requiring certain coverage can delay the drop in supply and surge in prices that typically follow a large loss. It has also shown that insurance premiums can be reduced by shifting some costs to taxpayers and to policyholders exposed to the smallest amount of risk. But those actions drive private suppliers out of the market, tend to increase losses, and may lead to higher prices over the long term.

- Since September 11, the supply of private insurance against terrorism has grown, prices have fallen, and coverage has become less restrictive. Experience with natural disasters strongly suggests that this adjustment will continue. However, private insurers may not be able to pay claims and continue issuing coverage after every contingency. Some imaginable events could produce losses that were bigger than the industry could afford. Also, another loss of unexpected size could occur, temporarily halting insurers' willingness and ability to issue coverage.

- Under current proposals for federal reinsurance of terrorism risks, the government would initially pay for most of the losses from a terrorist attack. Insurers would not pay any premiums up front for that coverage. However, CBO estimates that the government would need to charge insurers about $3 billion annually to fully compensate taxpayers for the risks they would assume under one of the proposals.

- As an alternative to providing reinsurance, the Congress could consider other measures to encourage the private sector to supply reinsurance following catastrophic events. Those measures include offering property owners incentives to mitigate risks, reducing federal assistance after disasters, changing the tax treatment of loss reserves held by insurers, and limiting damage awards.

# 1

# The Market for Property Insurance

The insurance market is usually divided into four parts: life, health, marine, and property. Property and casualty (P&C) insurers pay for losses of or damage to homes, commercial buildings, motor vehicles and airplanes, business equipment, and personal and business income from causes including fire, wind, explosion, and negligence. The terrorist attacks of September 11 hit P&C insurers harder than other insurers. Payment of claims for economic losses on that day will surpass $30 billion and could top $50 billion. Those losses exceed the combined total of insured losses from what had been the two largest single disasters in the United States—Hurricane Andrew, which struck south Florida in 1992, and the Northridge earthquake, which shook the greater Los Angeles area in 1994.

Property and casualty insurers do not sell "disaster" insurance. Rather, they cover the insured economic losses of individuals, businesses, and nonprofit organizations. A single claimant's losses would almost never amount to a national disaster, nor even a financial disaster, for a well-managed insurer. Indeed, the insurance market operates by pooling individual risks across a large and diverse population. Each insured property owner is protected against uncertain loss for the price of a pro rata share of the expected losses, which is paid into a fund and used to compensate the few owners who do suffer a loss. Thus, under normal circumstances, property insurers use annual premiums (and the investment income earned on reserves, retained earnings, and capital) to cover losses that would otherwise be personal catastrophes but not national disasters. Insurance disasters occur when a single event imposes losses on a large number of insured properties.[1] Occurrences of that magnitude are rare, which is why they are designated as happening, on average, once every 50 years (a 50-year loss) or once every 100 years (a 100-year loss).

Although P&C insurers have traditionally focused on natural disasters, such as hurricanes and earthquakes, manmade disasters also pose a large-scale threat to property. The challenge to insurers in providing disaster coverage is predicting the frequency and severity of any catastrophic event. Insurers have information and methods to make satisfactory predictions for natural disasters and can design and price insurance products to cover damage from those events. Understanding the risk of loss from willful acts is far harder because there is little basis for predicting the frequency, kind, and severity of future attacks.

Thus, when a new category of risk emerges, such as the threat of terrorist attacks, P&C insurers may issue coverage that excludes such losses, at least until insurers obtain more information and refine their methods of analysis and prediction. Prior to September 11, terrorism insurance was provided only implicitly; that is, most existing policies did not explicitly include or exclude losses from terrorism. After the attacks on the World Trade Center, insurers began to exclude terrorism from coverage under new policies and renewals.

---

1. See Christian Gollier, "Insurability" (paper presented at the National Bureau of Economic Research Conference on Insurance, Cambridge, Mass., February 1, 2002).

Overall, the market for property and casualty insurance is highly competitive and adaptive to new circumstances. It is also tightly regulated, especially for personal lines of insurance, such as homeowners and automobile. Traditionally, P&C insurers have experienced periodic increases and decreases in net income from underwriting activities.[2] However, the industry has always recovered from periods of heavy losses. A useful approach to understanding the market's performance is to examine the principal forces that affect it—supply, demand, and regulation.

## Supply

More than 100 companies sell property and casualty insurance in the United States. The names of the largest national operators—State Farm, Allstate, and American International Group—are familiar to most consumers. But many insurers specialize regionally.[3] Also, some states sponsor insurance programs, especially to cover high-risk properties.

The market for P&C insurance appears to be competitive nationally, with the three largest companies dividing a 23 percent market share and the top 10 firms collecting 44 percent of total premiums in 2000.[4] The industry relies heavily on information and capital, needing the former to assess risk and the latter to ensure the capacity to pay claims.

### Risk Assessment and Capital Holdings

Perhaps the easiest part of providing insurance is when a company's agent says to a property owner, "You're covered." For an insurer, agreeing to a binding contract to protect against specified losses is the final step in a process that includes establishing premiums that are actuarially sound and obtaining sufficient capital to pay claims. Insurance premiums must cover insured losses, administrative expenses, and a return on the capital that would be at risk if losses exceeded annual premiums and investment income.

To set premiums that are sound, insurers must allow for the uncertainty of the "true" distribution of future losses, especially for infrequent events that result in large losses. Even though many hurricanes, earthquakes, and man-made acts of destruction occur each year, catastrophes are rare. In the past century, only three of the most powerful hurricanes (Category 5 on the Saffir-Simpson scale, which measures hurricane strength on the basis of wind speed) have made landfall in the United States. Similarly, only four earthquakes exceeding 8.0 on the Richter scale are known to have struck the United States, and none affected a densely populated area.[5] (Of course, less powerful earthquakes and hurricanes can cause extensive damage and loss of life if they strike an area that is densely populated.) And although a number of incidents of terrorism have occurred in the United States in recent years, none approached the attacks of September 11 in magnitude or in severity. With such infrequent catastrophes, insurers cannot be sure that they have reliably estimated expected losses.[6] Moreover, the underlying probabilities of catastrophic events can change as a result of variations in climate, geology, and the global political environment.[7]

Because insurance losses may vary widely, insurers must hold sufficient capital reserves to bridge the gap between

2. Congressional Budget Office, *The Economic Impact of a Solvency Crisis in the Insurance Industry* (April 1994).

3. Martin W. Grace and others, *Catastrophe Insurance: Supply, Demand, and Regulation*, Working Paper No. 01-38-pk (Philadelphia: Wharton Catastrophic Risk Management Project, December 2001).

4. Data are from A.M. Best Company as reported in Insurance Information Institute, *The Fact Book 2002* (New York: Insurance Information Institute, 2002), p. 8.

5. Letter from Paul G.J. O'Connell, Chairman, Catastrophe Work Group, American Academy of Actuaries, to the General Accounting Office, May 7, 1997.

6. Howard Kunreuther, "The Role of Insurance in Managing Extreme Events: Implications for Terrorism Coverage," *Business Economics*, National Association for Business Economics, April 2002.

7. In the extreme, some risks may have loss distributions that cannot be observed or identified. See Nomura Fixed Income Research, *How the Events of 9/11 Affect Thinking About Risk* (New York: Nomura Securities International, January 3, 2002); and Howard Kunreuther, Robin Hogarth, and Jacqueline Meszaros, "Insurer Ambiguity and Market Failure," *Journal of Risk and Uncertainty*, vol. 7, no. 1 (1993), pp. 71-87.

annual premiums and actual losses. When catastrophic events occur, they adversely affect many properties and people. Individual losses are thus "bunched" in space and time, or correlated with one another. The variance of an insurer's losses—the distribution around the mean—is higher for correlated than for uncorrelated risks and requires larger reserves.

The added cost of holding larger amounts of capital means that insurers must charge a higher premium for correlated risks than for independent risks.[8] That difference can be seen in an example of two insurance portfolios: one for hurricane coverage (correlated risks) and one for fire (independent risks).[9] The hurricane portfolio exposes the insurer to a $100 million loss once every 100 years, whereas losses from fire alternate between $0.8 million and $1.2 million each year. Annual premiums of $1 million would cover the expected costs of each hazard. But to ensure its ability to pay claims every year, a fire insurer would need to hold only about $200,000 in capital reserves for those years in which losses were $1.2 million. An insurer covering losses from hurricanes, in contrast, would need to hold capital of $99 million for the $100 million loss that would occur, on average, once every 100 years. (Bond-rating agencies generally require that insurers have enough capital to pay off at least a 100-year loss. To attain a top rating, insurers may need to hold capital against a 250-year loss.)

With either portfolio, insurers would expect to lose a certain amount of capital each year, and that loss must be incorporated in the price of the insurance. The price must also reflect other costs to capital reserves. One such cost is the competitive return that investors must earn on their capital. The cost of capital is the opportunity cost of not investing the money in the most profitable alternative activity (after adjusting for risk). Insurers invest most of their capital in relatively safe, liquid assets (such as bonds) to assure policyholders and rating agencies that they can pay claims in the event of a disaster.

Taxes represent another cost to capital reserves. The interest on the bonds held by an insurance company is subject to corporate income taxes, which reduce the earnings available for return to the owners who provided the capital.[10] That taxation pushes up the premium that insurers must charge on catastrophic risk. If taxes absorb one-fourth of investment income, then an insurer's 8 percent market return on investments is reduced to 6 percent after taxes. Because shareholders could instead invest directly in a mutual fund that returned 8 percent, insurers would need to collect an additional 2 percentage points in premiums ($1.98 million in the previous example) to compensate providers of capital for hurricane exposure.[11] For fire risks, however, insurers would need to collect much less (only $4,000) to compensate stockholders for the cost of their capital. Thus, an insurer would need to charge policyholders nearly $3 million per year for the hurricane insurance but just over $1 million for the fire

8. Howard Kunreuther, "Insurability Conditions and the Supply of Coverage," in Kunreuther and Richard J. Roth Sr., eds., *Paying the Price: The Status and Role of Insurance Against Natural Disasters in the United States* (Washington, D.C.: Joseph Henry Press, 1998), pp. 37-38. Prices for catastrophe coverage must include a markup (or load) over expected losses even when competition is strong. Consequently, high prices for catastrophe coverage should not necessarily be interpreted as a market failure. See George Zanjani, "Pricing and Capital Allocation in Catastrophe Insurance" (working paper, Federal Reserve Bank of New York, May 21, 2001).

9. This example is based on one presented in Rade T. Musulin, "Can Our Children Afford 'Affordable' Insurance?" *Emphasis*, Tillinghast-Towers Perrin (1997), pp. 2-5.

10. Scott E. Harrington and Greg Niehaus, "Government Insurance, Tax Policy, and the Affordability and Availability of Catastrophe Insurance," *Journal of Insurance Regulation*, vol. 19, no. 4 (Summer 2001), pp. 591-612.

11. The restrictions precluding some risky investments do not represent a significant cost to insurers or investors. As long as investors get a fair risk-adjusted return on their investment, they do not need to receive an equity return on what is essentially a bond fund. An investor who chose a bond mutual fund would be satisfied with the returns passed through by an insurance company if it were not for the unfavorable tax treatment those returns received. Moreover, investors may not need to receive a higher return on their capital because of the risk of insurance losses. Risks that are not correlated with the rest of the market can be reduced through diversification and thus do not require a premium. Catastrophes such as hurricanes have very low correlation with the market and, hence, do not represent priced risk. In practice, investors might require some additional compensation for the great uncertainty of disasters, for example, or for the possibility of fraud.

insurance, even though both portfolios have the same expected overall loss.

To pay claims in the event of a disaster, insurers have options other than holding costly reserves. One option is to borrow the money, but that may not be possible because policyholders are not contractually obligated to purchase multiyear policies. If they were, their future premiums might serve as collateral. Also, borrowing may become more difficult after a disaster because insurers have the option of declaring bankruptcy.

Insurers can substantially reduce their cost of capital through risk diversification. By combining insurance pools whose losses have different causes and hence are likely to occur at different times (hurricanes in Florida, earthquakes in Japan, and terrorism in Europe, for example), insurers can narrow the gap between overall expected losses and the maximum probable loss in any one year, reducing their need for capital. Diversification can occur through a single insurer's underwriting different hazards (automobile, fire, and workers' compensation, for instance) in various locations. Alternatively, it can happen through reinsurance, a process in which one insurer sells some of its business and the associated income from premiums to another insurer. Through such sales, even specialized regional insurance companies can reduce their risks and need for capital.

Most reinsurers are large, well-capitalized companies. By operating in international markets, they are able to raise large sums of capital and diversify risks broadly. The four largest reinsurers in 2000 were Munich Re, Swiss Re, General Electric Global Insurance Holdings, and Berkshire Hathaway Reinsurance Group.

**Industry Earnings**

Insurers' capacity to pay claims depends on their net worth. And insurers' net worth hinges at least in part on their ability to charge appropriate premiums so as to generate earnings and attract capital. In 2000, the U.S. property and casualty industry earned $9.5 billion in pretax operating income on $301 billion in net premiums (*see Table 1*). Earnings were down significantly from the peak years of the mid-1990s, a pattern that is consistent with the industry's so-called underwriting cycle (several years of rising earnings followed by several years of falling earnings).[12] For 2001, the industry reported its first overall loss since 1992.

The 1990s was a period of volatile underwriting losses. Although a certain level of underwriting losses is normal for insurers, they will lose money and may become insolvent if losses grow too large. But the 1990s also saw the property and casualty industry add capital (also called policyholders' surplus) at a rapid rate; the ratio of capital to premiums nearly doubled between 1990 and 1998. Since then, capital holdings have fallen by about $45 billion, primarily because of the high level of underwriting losses and declining returns on investments. At the end of 2001, domestic insurers' net worth stood at $290 billion.[13] The surplus for reinsurers was similarly large; at the end of 2000, the top 25 global reinsurers, with more than 80 percent of the market, had $208 billion of surplus.[14] Industrywide measures of insurers' ability to pay can be misleading, however, if claims from disasters are concentrated in relatively few firms.[15]

---

12. Several explanations exist for the underwriting cycle. Some analysts emphasize capital market imperfections; see Kenneth A. Froot and Paul G.J. O'Connell, *On the Pricing of Intermediated Risks: Theory and Application to Catastrophe Reinsurance*, Working Paper No. 6011 (Cambridge, Mass.: National Bureau of Economic Research, April 1997). Other analysts are less certain of the causes of price fluctuations; see Anne Gron and Deborah Lucas, "External Financing and Insurance Cycles," in David F. Bradford, ed., *The Economics of Property-Casualty Insurance* (Chicago: University of Chicago Press, 1998), pp. 5-27.

13. As of March 31, 2002, capital holdings of domestic reinsurers were $25.3 billion, a drop of just $1.3 billion since June 30, 2001. For data on reinsurers, see the statement of Franklin W. Nutter, President, Reinsurance Association of America, before the Senate Commerce, Science, and Transportation Committee, October 30, 2001; updated data are available at www.reinsurance.org.

14. Those figures also include the companies' primary insurance operations. See Standard & Poor's, "Reinsurance Outlook 2002: Price Surge Bullish for Earnings," Commentary in *Ratings Direct*, December 18, 2001.

15. J. David Cummins, Neil Doherty, and Anita Lo, *Can Insurers Pay for the 'Big One'? Measuring the Capacity of the Insurance Industry to Respond to Catastrophic Losses*, Working Paper No. 98-11-B (Philadelphia: Wharton Financial Institutions Center, June 24, 1999), available at http://fic.wharton.upenn.edu/fic/papers/98/9811.pdf. Also see General Accounting Office, *Insurers' Ability to Pay Catastrophe Claims*, GAO/GGD-00-57R (February 8, 2000).

**Table 1.**

## Operating Results for U.S. Property and Casualty Insurers, 1990 Through 2001

(In millions of dollars)

| Year | Net Premiums | Net Underwriting Loss[a] | Pretax Operating Income[b] | Policyholders' Surplus[c] |
|---|---|---|---|---|
| 1990 | 218,100 | -21,652 | 11,249 | 138,401 |
| 1991 | 223,243 | -20,458 | 13,789 | 158,658 |
| 1992 | 227,751 | -36,260 | -2,526 | 163,080 |
| 1993 | 241,691 | -18,094 | 14,551 | 182,275 |
| 1994 | 250,709 | -22,083 | 11,604 | 193,346 |
| 1995 | 259,803 | -17,375 | 19,459 | 230,001 |
| 1996 | 268,730 | -17,162 | 20,801 | 255,527 |
| 1997 | 276,568 | -6,030 | 35,469 | 308,479 |
| 1998 | 281,621 | -16,572 | 23,354 | 333,327 |
| 1999 | 286,934 | -24,429 | 14,426 | 334,348 |
| 2000 | 301,000 | -32,300 | 9,500 | 319,000 |
| 2001 | 323,977 | -52,990 | -15,164 | 289,649 |

Source: Congressional Budget Office based on A.M. Best Company, *Best's Aggregates & Averages: Property-Casualty* (Oldwick, N.J.: A.M. Best Company, 2000); A.M. Best Company, "2000 Property/Casualty Results: Insurers Not Out of the Woods" (Oldwick, N.J.: A.M. Best Company, April 16, 2001) for 2000 data; and Insurance Services Office, Inc., "Property/Casualty Industry Suffers First-Ever Net Loss in 2001; Surplus Drops as Terrorist Attack and Poor Investment Results Pummel Earnings" (press release, Jersey City, N.J., April 15, 2002), available at www.iso.com (for 2001 data).

Note: Data are for private insurance companies, so they exclude state funds.

a. Includes payments of dividends to policyholders. In 2001, dividends were $2.3 billion. See Ruth Gastel, ed., "Financial and Market Conditions," in *Insurance Issues Update* (New York: Insurance Information Institute, May 2002).

b. Excludes the realization of capital gains or losses.

c. Insurers' capacity to issue insurance depends on their net worth (assets minus liabilities), or capital and surplus. The largest component of net worth is surplus, which is an insurer's accumulated stock of retained earnings. Capital is shareholders' equity of a publicly owned insurance company.

### State-Sponsored Programs

A few states, notably Florida and California, actively supply property and casualty insurance (*see Table 2*). Florida has established three programs: a residential insurer of last resort; an insurer that provides only hurricane coverage; and a reinsurer. Following the Northridge earthquake, California created the California Earthquake Authority (CEA) to issue earthquake insurance to homeowners. And Hurricane Iniki in 1992 prompted Hawaii to establish the Hawaii Hurricane Relief Fund to sell insurance that covered hurricane damage.[16] In addition, the states of Alabama, Florida, Louisiana, Mississippi, North Carolina, South Carolina, and Texas have initiated "beach plans," risk-sharing pools specifically designed to provide homeowners' insurance in coastal areas. (See Appendix A for a more complete discussion of state-sponsored insurance programs in Florida and California.)

Rates in Florida's programs are generally set below market levels and at costs lower than expected losses, whereas rates in California typically reflect expected losses.[17] All insurers operating in Florida are residually liable for fund-

16. The recovery of the private market for hurricane insurance enabled the Hawaii Fund to cease issuing and renewing policies as of December 1, 2000.

17. Personal communication from Richard J. Roth Jr., a consulting actuary, to the Congressional Budget Office, January 8, 2002. Although the California Earthquake Authority's rates may be actuarially sound overall, critics charge that rates for the riskiest properties are capped to promote affordability. In addition, unlike private companies, the CEA imposes no load for the cost of its capital. See Tillinghast-Towers Perrin, "CEA Consulting Team Report" (New York, July 5, 2001), pp. 65-73.

**Table 2.**

## Selected State-Sponsored Insurance Suppliers

| State-Sponsored Insurer | Type of Coverage | Payment Capacity (Billions of dollars) |
|---|---|---|
| California Earthquake Authority | Supplemental earthquake coverage | 7.36[a] |
| Florida Hurricane Catastrophe Fund | Reinsurance for hurricane losses | 11.0[b] |
| Florida Residential Property and Casualty Joint Underwriting Association | Standard homeowners' coverage | 1.86 |
| Florida Windstorm Underwriting Association | Coverage for residential losses from wind perils | 5.8 |
| Hawaii Hurricane Relief Fund | Hurricane coverage for commercial and residential buildings | 1.49[c] |

Source: Congressional Budget Office based on information from the insurance suppliers.

a. As of November 30, 2001.

b. As of December 31, 2001. The payment capacity to cover the first hurricane season is $11 billion. The payment capacity for the following year's hurricane season would be $7.9 billion.

c. Should claims exceed this sum, the Hawaii Hurricane Relief Fund can impose assessments on all property and casualty premiums written in the state to raise additional funds.

ing future deficits incurred by the state funds, which appear to be undercapitalized. Such losses would be passed on to policyholders through postdisaster assessments. Thus, the current low prices of the Florida programs do not represent the full cost of the insurance.

Some state-sponsored funds, including the CEA and the Florida Hurricane Catastrophe Fund, also enjoy significant tax advantages, enabling them to charge lower prices. Florida's other two state-sponsored funds (the Florida Windstorm Underwriting Association and the Florida Residential Property and Casualty Joint Underwriting Association) are being restructured and merged into a single fund to qualify for tax-exempt status. The tax exemptions shift costs from the states to federal taxpayers.

None of the state-sponsored insurers has been tested by a disaster on the scale of Hurricane Andrew, although some have paid out claims for smaller events. Hurricanes that struck Florida in 1994, 1995, and 1998 led to several hundred million dollars in disbursements by that state's sponsored insurers. Those payments were funded by assessments imposed on private insurers and policyholders. The CEA has paid out only 695 claims, totaling roughly $473,000, since its inception.[18]

States sometimes require private insurers to provide disaster coverage. California has required all firms offering homeowners' insurance in the state to also provide earthquake coverage at the option of the buyer, for example, since 1985.

## Demand

Consumers buy property and casualty insurance for a variety of reasons. Many property owners prefer the certainty of a fixed annual premium to the risk of uninsured catastrophic loss. Others with mortgages buy coverage because lenders require insurance on assets used as collateral. Most consumers' insurance needs are met through standard homeowners' policies, which cover losses from hurricanes but not from floods or earthquakes. Flood insurance is available from the federal government, and its purchase is mandatory in most high-risk areas. Mortgage

18. Personal communication from Stan Devereux, Legislative and Public Affairs Director, California Earthquake Authority, February 13, 2002.

lenders rarely require earthquake insurance, but some households insure against quakes through a separate policy option.[19]

Self-insurance plans are the most widespread substitutes for purchased P&C insurance. Many large businesses, nonprofit organizations, and state and local governments with numerous and diversified risks choose to forgo coverage from an insurer and instead absorb losses as they occur. The owners of a large commercial property that is a landmark can self-insure by spreading ownership among several million stockholders. Indeed, they may have little practical choice but to do so. Individuals and small businesses can partly self-insure by purchasing insurance with high deductibles or coinsurance.

Consumers increasingly decide to self-insure as premiums for purchased insurance rise and as coverage terms become more restrictive. In California, complaints persist that prices for earthquake insurance sold through the state-sponsored program are too high. The program's large deductibles, numerous exclusions, and tight caps on contents and living expenses mean that homeowners with policies would bear substantial uninsured loss after a quake.[20] As a result, about 16 percent of California residences now have earthquake insurance, down from 30 percent before the Northridge earthquake.[21] However, homeowners' willingness to pay for disaster insurance may be affected by misperceptions about the frequency of rare events and the availability of federal aid after a disaster.[22]

Most state and local governments do not buy P&C insurance, especially for infrastructure such as streets and highways, bridges, and public buildings. Separate coverage may be unnecessary because the Federal Emergency Management Agency provides state and local governments with public assistance grants to repair infrastructure that is damaged from a disaster.

## Regulation

Most states regulate premiums for property and casualty insurance, including homeowners' and automobile coverage, that is sold to individuals. The goal of price regulation is to keep insurance affordable. During the 1990s, requests by Florida's insurers to increase premiums for homeowners' insurance were often denied, delayed by the appeals process, or granted only in part.[23] As a result, rates for homeowners' insurance in Florida are at or below rates in nearby coastal states, even though the risk of hurricane-related losses is higher in Florida.[24]

---

19. There is an exception: Freddie Mac requires earthquake insurance for some condominiums in California. Apparently, single-family homes are more resistant to earthquake damage than are multifamily properties. Also, any losses from the fires that often accompany earthquakes are covered under standard homeowners' policies. See Morgan Stanley Dean Witter, "Fannie Mae: Tremble Not Over Trembler Risk" (New York: Morgan Stanley Dean Witter, August 9, 2001).

20. For instance, there is typically a 15 percent deductible of the value of the property. Consequently, policyholders more than 20 miles from a fault line, and thus less exposed to losses, are unlikely to receive any reimbursement for claims. In addition, individuals may underestimate their exposure to low-probability, high-loss events. See Howard Kunreuther and Mark Pauly, *Ignoring Disaster: Don't Sweat the Big Stuff,* Working Paper No. 01-16-HK (Philadelphia: Wharton Risk Management and Decision Processes Center, October 11, 2001), available at http://grace.wharton.upenn.edu/risk/wp/wplist01.html.

21. Price increases may explain some of the drop in coverage rates. Some research indicates that consumers—at least those in Florida and New York—are more responsive to changes in price for catastrophe coverage than for noncatastrophe coverage. See Martin Grace, Robert W. Klein, and Paul R. Kleindorfer, *The Demand for Homeowners Insurance with Bundled Catastrophe Coverage,* Working Paper No. 02-06 (Philadelphia: Wharton Project on Managing and Financing Extreme Risks, January 2002).

22. See Risa Palm, "Demand for Disaster Insurance: Residential Coverage," in Kunreuther and Roth, eds., *Paying the Price,* pp. 51-66.

23. Robert Klein, "Regulation and Catastrophe Insurance," in Kunreuther and Roth, eds., *Paying the Price,* pp. 196-201.

24. Collins Center for Public Policy, *Final Report of the Academic Task Force on Hurricane Catastrophe Insurance* (Tallahassee, Fla.: Collins Center for Public Policy, 1995), pp. 43-46. Since 1995, rates for homeowners' insurance in Florida have approached those in other coastal states.

Price regulation can make coverage more affordable in the short run. But regulation reduces the availability of and boosts prices for private insurance in the long run by pricing private suppliers out of the market (they cannot compete as claims rise and premiums stay the same). Markets in which regulatory intervention has been consistently minimal, such as the market for commercial property insurance, have performed reasonably well over the long term. For example, as a share of property at risk, the market for commercial earthquake insurance in California, which is unregulated, is several times larger than the residential market, which is regulated. Primary insurers are better able to attract capital and reinsure commercial risks because they can charge higher premiums.[25] Also contributing to the willingness of insurers to issue policies in the commercial market may be commercial structures' enhanced ability to withstand disasters, either because of their location or because of significant investments in loss prevention.

Terrorism insurance and other commercial coverage is generally sold at unregulated market prices. For primary insurers, most states have approved some exclusions of coverage for terrorism losses. Primary insurers can also use surplus lines (coverage that is not available from insurers licensed or "admitted" by a state) to exclude terrorism coverage. In addition, because state insurance commissioners cannot compel reinsurers to issue coverage, many reinsurers will choose to exclude terrorism risk from their P&C policies. Some states, however, do not allow terrorism claims to be excluded from commercial P&C coverage.[26] Consequently, primary insurers in those states (New York, California, Florida, Georgia, and Texas) must weigh the risk of being overexposed to losses from terrorism against the loss of revenue from not issuing policies.

25. For detailed information on coverage and comparisons of residential and commercial markets prior to the creation of the CEA, see California Department of Insurance, *California Earthquake Zoning and Probable Maximum Loss Evaluation Program: An Analysis of Potential Insured Earthquake Losses from Questionnaires Submitted by Property/Casualty Insurers in California* (Los Angeles: California Department of Insurance, 1998), especially Tables 7 and 8, pp. 12-14.

26. Exclusions of terrorism coverage cannot be part of workers' compensation coverage. Fires that may follow a terrorist attack also must be covered under standard homeowners' policies in about 30 states. Terrorism coverage is not an issue in the residential market because houses are not considered likely targets. See American Academy of Actuaries, *Terrorism Insurance Coverage in the Aftermath of September 11th: A Public Statement by the Extreme Events Committee of the American Academy of Actuaries* (April 17, 2002), available at www.actuary.org/pdf/casualty/terrorism_17apr02.pdf; and Standard & Poor's, "Commercial Insurers to Cut Coverage for Losses Caused by Terror Attacks," *Ratings Direct*, January 9, 2002. As of January 2002, blanket exclusions of terrorism risk were not being imposed on small business and personal lines of insurance. Instead, exclusions for those lines cover only losses from nuclear, chemical, or biological terrorism. Personal communication from Rade T. Musulin, Vice President, Florida Farm Bureau Insurance Companies, January 18, 2002.

# CHAPTER 2

# The Insurance Market's Response to Disasters

Both natural and manmade disasters trigger a predictable sequence of financial events for insurers and reinsurers. In the short run, big and unanticipated losses lead to large insurance payouts, which reduce insurers' capital reserves and heighten their uncertainty about losses from future events. Those effects lessen insurers' ability and willingness to provide coverage at existing prices. As the supply of insurance dwindles, sometimes simultaneously with an increase in demand, prices climb. Higher premiums in turn raise concerns about affordability. In the long run, the higher premiums will attract new capital and spur urgent assessments of unfamiliar risks. As a result, supply will begin to recover; prices will stabilize and then fall, although not necessarily to pre-disaster levels because of insurers' higher perceived risk of loss.

## Recent Unanticipated Losses

Hurricane Andrew, the Northridge earthquake, and the terrorist attacks of September 11 resulted in major losses for property and casualty insurers.

The destruction from Hurricane Andrew—which swept ashore 20 miles south of Miami on August 24, 1992—set a new record for insured losses. Property damage from the Category 5 hurricane totaled $34 billion (in 1999 dollars), $17 billion of which was insured (*see Table 3*).[1] Almost two-thirds of the dollar amount of all claims, approximately $11 billion, was paid to holders of homeowners' policies.[2] Commercial policies accounted for most of the remainder.

On January 17, 1994, an earthquake measuring 6.7 on the Richter scale occurred at a previously unknown fault beneath California's San Fernando Valley, 20 miles northwest of downtown Los Angeles.[3] Peak ground movement during the 15-second quake was among the most rapid ever recorded in the United States. Damage from the Northridge earthquake totaled $43 billion (in 1999 dollars), $16.6 billion of which was insured. Claims on homeowners' policies constituted more than three-

1. The National Hurricane Center recently changed the classification of Hurricane Andrew from Category 4 to Category 5. Scientists determined that maximum sustained wind speeds during the hurricane were 165 miles per hour—or 20 miles per hour faster than they had previously estimated. See National Oceanic and Atmospheric Administration, "After 10 Years, Hurricane Andrew Gains Strength" (press release, Washington, D.C., August 21, 2002), available at www.nhc.noaa.gov.

2. Eugene Lecomte and Karen Gahagan, "Hurricane Insurance Protection in Florida," in Howard Kunreuther and Richard J. Roth Sr., eds., *Paying the Price: The Status and Role of Insurance Against Natural Disasters in the United States* (Washington, D.C.: Joseph Henry Press, 1998), p. 105.

3. Richard J. Roth Jr., "Earthquake Insurance Protection in California," in Kunreuther and Roth, eds., *Paying the Price*, pp. 67-95.

**Table 3.**

## Insured Losses from Hurricane Andrew

| Type of Insurance | Insured Losses (Millions of 1999 dollars) | Percentage of Total Insured Losses | Number of Direct Claims |
|---|---|---|---|
| Homeowners | 11,111 | 65.0 | 280,000 |
| Commercial Multiperil | 3,767 | 22.0 | 50,517 |
| Fire and Allied Lines (Commercial)[a] | 1,062 | 6.2 | 24,467 |
| Automobile, Physical Damage | 365 | 2.1 | 161,400 |
| Mobile-Home Owners | 204 | 1.2 | 11,779 |
| Farmowners | 16 | 0.1 | 1,245 |
| Other Types[b] | 559 | 3.3 | 17,177 |
| **Total** | **17,084** | **100.0** | **546,585** |

Source: Congressional Budget Office based on data from the Florida Department of Insurance, cited in Eugene Lecomte and Karen Gahagan, "Hurricane Insurance Protection in Florida," in Howard Kunreuther and Richard J. Roth Sr., eds., *Paying the Price: The Status and Role of Insurance Against Natural Disasters in the United States* (Washington, D.C.: Joseph Henry Press, 1998), p. 105.

Note: Numbers may not add up to totals because of rounding.

a. Includes coverage for losses caused by fire, lightning, wind and water damage, and vandalism.

b. Includes coverage for boat owners, as well as other insurance lines.

quarters of the total dollar value of insured claims.[4] Claims might have been far higher, but only 40 percent of homeowners in the Northridge area carried earthquake coverage.[5]

The September 11, 2001, terrorist attacks on the World Trade Center completely destroyed the New York City landmark. Financial losses from the attacks will be the largest ever for a single event in the United States. Current estimates of insured losses exceed $30 billion.[6] Prior to the attack, insurers had collected almost no premiums for terrorism risks. Moreover, none of the insurers anticipated any scenario that would result in the complete destruction of the 110-story twin towers. In setting commercial policy premiums, insurers largely ignored the February 1993 bombing of the World Trade Center, even though that event caused roughly $725 million in insured property losses (in current dollars), 1,000 injuries, and six deaths.

4. The Northridge earthquake was unusual in that most of its insured losses were residential rather than commercial. (About 80 percent of the insured coverage in California today is commercial coverage.) After the earthquake, the California Department of Insurance raised the probable maximum loss for residences in the Los Angeles and San Francisco areas by 25 percent but made no changes to the probable maximum loss for commercial structures. See California Department of Insurance, *California Earthquake Zoning and Probable Maximum Loss Evaluation Program: An Analysis of Potential Insured Earthquake Losses from Questionnaires Submitted by Property/Casualty Insurers in California* (Los Angeles: California Department of Insurance, 1998).

5. Insurance Services Office, *Catastrophes: Insurance Issues Surrounding the Northridge Earthquake and Other Natural Disasters* (New York: Insurance Services Office, 1994), p. 14.

6. Statement of Matthew C. Mosher, Group Vice President, A.M. Best Company, before the House Committee on Financial Services, September 26, 2001. As of October 2, 2001, industry estimates of insured losses from the terrorist attacks ranged from $30 billion to $70 billion. See David Pilla, "The Cost of Terror," *Best's Review*, A.M. Best Company (October 2001). Business-interruption claims are still being decided, and concerns about asbestos-related claims—possibly totaling as much as $7 billion—have been voiced. However, the loss of life from the attacks is less than was originally feared, which reduces claims for life insurance and workers' compensation. See Standard & Poor's, "Commercial Lines Outlook for 2002: Negative Fundamentals Persist Despite Surging Prices," *Ratings Direct*, January 31, 2002.

## Short-Run Market Adjustments

Andrew, Northridge, and September 11 each set off a "supply shock" in insurance markets. Unanticipated losses from major disasters reduce insurers' net worth and increase their uncertainty about the level and variance of future losses. Both factors tend to rein in the supply of coverage and elevate prices in the short term.

### Reduced Income and Capital

After each of the three disasters, the affected property and casualty insurers saw their earnings fall and their capital initially drop in value. Of the two natural disasters, Hurricane Andrew was the more damaging to the industry. The year-over-year increase in underwriting losses nearly doubled in 1992, the year that Andrew struck, and the industry recorded its first operating loss of the decade (*see Table 1* on page 5). Three insurers each paid claims in excess of $1 billion, and 11 small insurance companies filed for bankruptcy.[7] (The Florida State Guaranty Fund paid claims made against those insolvent firms only up to the $300,000 statutory limit.)

Despite the magnitude of the industry's financial losses, however, capital grew by more than $4 billion in 1992. That anomalous result in part reflects the weakness of overall industry data on capital, which fail to show the wide dispersion of capital across the industry. While some firms were collapsing into insolvency, others were profitably able to attract new sources of revenue. That result also shows that the flow of capital into the industry is not merely the remainder left after expenses are paid and dividends are distributed. Capital flows are strongly affected by investors' perceptions of future rates of return; likewise, earnings are affected by the returns that insurers receive on the capital they invest. The unusually strong stock market of the 1990s bolstered insurers' income and capital during that decade.

The Northridge earthquake significantly drained the financial resources of many insurers, but none went bankrupt as a result. (Technically, one insurer, 20th Century, did become insolvent from Northridge quake claims. However, a capital infusion from its parent company enabled the firm to pay all its claims without activating the California Insurance Guarantee Fund.) Underwriting losses for the industry grew by just over 20 percent in 1994, but they do not reflect the full cost of the quake because many claims were paid over the following years as the full extent of damages became apparent.[8] Industry capital continued to grow in 1994 but at a slower rate.

Despite their severity, the financial shocks from Andrew and Northridge pale in comparison with the losses from September 11. That one day shattered the industry's expectations about maximum insured losses from a single event as annual earnings were wiped out and global capital was effectively reduced by more than $30 billion (before taxes). Two insurers—one in Denmark, the other in Japan—failed as a result of the attacks. According to Standard & Poor's, losses from the Trade Center bombings would have to exceed $50 billion before they threatened the solvency of the U.S. insurance system.[9] (The effect on domestic insurers is dampened somewhat because more than half of the total losses will probably be covered by reinsurers.) So far, no U.S. insurers or major international reinsurers have failed because of the attacks, but rating agencies have downgraded the credit of several insurance companies.[10]

7. U.S. Senate, Bipartisan Task Force on Funding Disaster Relief, *Report of the Senate Task Force on Funding Disaster Relief,* Document No. 104-4 (March 15, 1994), p. 31.

8. Insured losses from the Northridge quake rose from an initial estimate of $2.5 billion to $10.4 billion one year later. The last estimate (from 2000) was $15 billion. See Richard J. Roth Jr., "A Government Experiment: The California Earthquake Authority" (presentation given at the Sixth International Conference on Seismic Zonation, Palm Springs, Calif., November 15, 2000).

9. Losses above that level would hurt the credit quality of a number of companies. See Standard & Poor's, "World Trade Center Attack Will Not Cripple Insurance Industry," *Ratings Direct,* September 14, 2001.

10. Nonetheless, one rating agency also stated that the "overwhelming majority of these insurers and reinsurers are expected to maintain secure financial strength ratings." See Standard & Poor's, "S&P Rating Actions on Insurers with World Trade Center Exposures" (press release, New York, September 20, 2001).

### Higher Uncertainty

Before accepting a risk, an insurer must be confident that it understands the magnitude of the commitment and that it has priced the premiums to be commensurate with that risk. Large, unanticipated losses require insurers to reassess risks.

Before Hurricane Andrew, the largest insured loss from a tropical storm resulted from Hurricane Hugo, which struck Georgia and the Carolinas in 1989, causing about $4 billion in insured losses. After Hugo, insurers reassessed their potential risks and determined that future hurricane losses would not exceed $8 billion. Thus, the $17 billion in insured losses from Hurricane Andrew invalidated their forecasting models and prompted an urgent reassessment of the risk of loss. Those efforts were hampered by meteorologists' uncertainty as to whether Hurricane Andrew was a loss that would occur once every 50 years, once every 100 years, or even less frequently.

In the weeks immediately following September's terrorist attacks, many insurers admitted that they had no systematic method for assessing or pricing terrorism risk. Consequently, they had no means of providing such coverage in a way that was consistent with their fiduciary obligations to shareholders.[11]

### Decreased Coverage and Increased Premiums

After a disaster, insurers try to lessen their exposure to catastrophic risk by reducing coverage and pushing up premiums. Following Hurricane Andrew, for example, 39 insurers sought to cancel or refuse to renew 844,433 Florida homeowners' policies.[12] Reinsurers reduce their exposure to risk by raising the deductible and coinsurance amounts that primary insurers must pay and by limiting coverage. One primary insurer's catastrophe deductible skyrocketed from $30 million to $100 million after Hurricane Andrew.[13] After the Northridge earthquake, insurers representing 95 percent of the homeowners' insurance market in California ceased issuing policies or severely restricted their sales.[14] And since September 11, securing terrorism coverage has been difficult for some owners of commercial property. Rates to insure large commercial properties in the Northeast increased by more than 30 percent, on average, between October 1, 2001, and January 1, 2002, the point at which most reinsurance contracts expired.

One study attempted to quantify the effect of catastrophic losses on insurance supply (and prices) during the 1990s. It found that a $10 billion loss reduced the quantity of insurance by 5 percent to 16 percent and raised the average reinsurance price by 19 percent to 40 percent in the following year.[15]

Prices for reinsurance climb after disasters because the supply of insurance shrinks. After Hurricane Andrew, reinsurance prices nearly doubled (*see Figure 1*).[16] By 1995, the Andrew-induced run-up in reinsurance prices had ended. Subsequently, reinsurance prices fell by 30 percent, mostly between 1995 and 1998.[17] But prices rose sharply again after September 11. Standard & Poor's

11. Dwight Jaffee and Thomas Russell, "Extreme Events and the Market for Terrorist Insurance" (paper presented at the National Bureau of Economic Research Conference on Insurance, Cambridge, Mass., February 1, 2002).

12. Florida Department of Insurance, cited in Lecomte and Gahagan, "Hurricane Insurance Protection in Florida," p. 106.

13. Anne Gron and Andrew Winton, "Risk Overhang and Market Behavior," *Journal of Business*, vol. 74, no. 4 (October 2001), pp. 591-612.

14. Statement of Gregory Butler, Chief Executive Officer, California Earthquake Authority, before the House Committee on Banking, June 24, 1997.

15. Kenneth A. Froot and Paul G.J. O'Connell, *The Pricing of U.S. Catastrophe Reinsurance*, Working Paper No. 6043 (Cambridge, Mass.: National Bureau of Economic Research, May 1997).

16. Kenneth A. Froot, "Introduction," in Froot, ed., *The Financing of Catastrophe Risk* (Chicago: University of Chicago Press, 1999), p. 9; and Guy Carpenter & Company, Inc., "Global Reinsurance Analysis 1998: A Guy Carpenter Special Report" (New York, Guy Carpenter, September 1998).

17. The reinsurance price index adjusts for changes in relative prices. See Paragon Reinsurance Risk Management Services, *Paragon Catastrophe Price Index* (Minneapolis, Minn.: Paragon Reinsurance Risk Management Services, January 2001).

**Figure 1.**

## Catastrophe Reinsurance Price Index, 1984 to 2001

(1984 = 1.00)

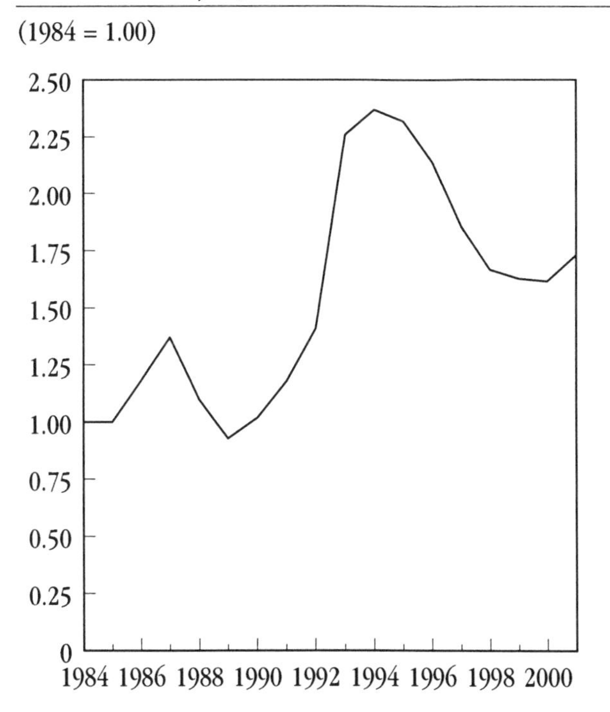


Source: Congressional Budget Office based on data from Paragon Reinsurance Risk Management Services.

Note: The index shows changes in the price of catastrophe reinsurance relative to a base of one. Data points are from January of each year. In July 1994, the index reached its peak of 2.47.

expects reinsurance prices to increase by 25 percent to 35 percent in 2002; that price hike will be accompanied by lower limits on payouts per event and exclusions for losses related to terrorism.[18]

## Long-Run Market Adjustments

After a disaster strikes, the insurance industry enters a period of adjustment. During that adjustment period, which varies in length after each disaster, higher prices for insurance attract additional investment into the disrupted market. As a result of that infusion of capital, capacity recovers, risks are reassessed and better understood, and prices typically decline—although not to predisaster levels. Instead, prices may remain permanently elevated because disasters can increase insurers' perceived risks of the probability of future disasters and their potential costs. For example, between 1993 and 1997, premiums for private earthquake policies went up by 63 percent in Los Angeles in real (inflation-adjusted) terms and by 57 percent in San Francisco (*see Table 4*). As of 2000, rates had risen by just 1 percent in Los Angeles while remaining unchanged in San Francisco. Although the industry's adjustment to the financial and forecasting shocks from Andrew and Northridge is largely complete, its adjustment to the shocks from September 11 is still under way.

Following each of those three disasters, capital flowed into the insurance industry from investors attracted by the prospect of high rates of return. Those capital flows largely eliminated the drop in policyholders' surplus, or net worth, for most firms paying large claims. They also supported the expansion of the industry's capacity. In addition, expectations of high rates of return led to several innovations in capital markets, including catastrophe bonds and options, which have provided about $2.5 billion of additional capacity per year (see Appendix B for details).

Capital holdings nearly doubled between Hurricane Andrew and the World Trade Center attacks, climbing from $163 billion as of December 31, 1992, to over $300 billion nearly a decade later.[19] Since September 11, insurers have raised substantial amounts of new capital—up to $25 billion, according to some estimates.[20] Much of that

18. Standard & Poor's, "Reinsurance Outlook 2002: Price Surge Bullish for Earnings," *Ratings Direct*, December 18, 2001.

19. Data from insurance-rating company A.M. Best indicate that as of March 31, 2001, the industry's surplus was $302 billion, compared with $333 billion one year earlier. Unrealized losses on common stock were the primary reason for the drop in the surplus. See A.M. Best Company, "Property/Casualty Premiums Show Strong Growth, Earnings Remain Depressed" (Oldwick, N.J.: A.M. Best Company), July 2, 2001.

20. See Christopher Oster, "Insurance Companies Benefit from Sept. 11, Still Seek Federal Aid," *Wall Street Journal*, November 15, 2001, p. 1; and statement of Kenneth A. Froot, Professor of Finance, Harvard Business School, before the Senate Committee on Banking, Housing, and Urban Affairs, October 24, 2001. The latest estimate is from Standard & Poor's, "Reinsurance Outlook 2002."

**Table 4.**

## Annual Premiums for a $150,000 Frame Home in California, 1993 Through 2000

(In 2000 dollars)

| Year | Homeowners' Policy | Earthquake Policy (Private)[a] | Earthquake Policy (CEA)[b] | Total Cost of Private Insurance |
|---|---|---|---|---|
| | | **Los Angeles—Central City** | | |
| 1993 | 810 | 350 | n.a. | 1,160 |
| 1996 | 725 | 470 | n.a. | 1,195 |
| 1997 | 840 | 570 | 400 | 1,410 |
| 2000 | 730 | 575 | 360 | 1,305 |
| | | **San Francisco** | | |
| 1993 | 690 | 380 | n.a. | 1,070 |
| 1996 | 675 | 465 | n.a. | 1,140 |
| 1997 | 710 | 595 | 625 | 1,305 |
| 2000 | 670 | 595 | 560 | 1,265 |

Source: Congressional Budget Office based on information from California Department of Insurance, Statistical Analysis Bureau, *Homeowners Premium Survey*, for 1993, 1996, and 1997. Data for 2000 came from the California Department of Insurance's Web site at www.insurance.ca.gov/docs/FS-Looking.htm.

Notes: n.a. = not applicable.

The Los Angeles example covers zip code 90011; the San Francisco example covers zip code 94109 (except for earthquake policy rates in 2000, which are based on the city's Nob Hill area). Minor inconsistencies occurred in the examples about which the surveys requested information. For instance, the 1993 survey used a home built in 1985, whereas the 1996 survey used one built in 1990.

Rates charged by the California Earthquake Authority (CEA) vary with a home's age. Rates for a frame home built after 1991 in the Nob Hill area of San Francisco are $3.30 per $1,000 of coverage, while rates for a home built prior to 1978 are $4.80 per $1,000. For homes built between 1979 and 1990, the rate is $3.75 per $1,000 of coverage, before adjusting for inflation, which is the rate used above.

a. The private rates cited are the average rates; premiums varied considerably among companies. (In the private market, brick homes are up to four times more expensive to insure against earthquakes.)

b. Coverage through the California Earthquake Authority is not as extensive as private coverage, so cost comparisons are problematic.

capital has been generated by new firms that have been incorporated in Bermuda to take advantage of that island's favorable tax and regulatory treatment. The rapid inflow of capital suggests that the current spike in rates for commercial property and casualty insurance may end more quickly than it has after previous disasters.[21] However, according to bond-rating agencies, most of the new capital is not being used to back terrorism risks because of insurers' difficulty in pricing those risks.

## States' Responses to Reduced Supply and Higher Prices for Coverage

States have generally reacted to catastrophic events in three ways: by requiring insurers to continue supplying coverage, by holding down rates, and by creating state-sponsored insurers to provide coverage at below-market rates. For example, faced with the withdrawal of insurers from the market following Hurricane Andrew, the Florida Department of Insurance required insurers operating in the state to renew most of their existing policies. Specifically, in any single year, insurers were permitted to drop no more than 10 percent of their residential policies in a county and no more than 5 percent statewide.[22]

21. Letter from Warren E. Buffett, Chairman of the Board, Berkshire Hathaway, to the Shareholders of Berkshire Hathaway, November 9, 2001, available at www.berkshirehathaway.com.

22. Collins Center for Public Policy, *Final Report of the Academic Task Force on Hurricane Catastrophe Insurance* (Tallahassee, Fla.: Collins Center for Public Policy, 1995); and Florida Insurance Council,

Following the Northridge earthquake, providers of homeowners' insurance sought to withdraw from California to avoid that state's mandate to also offer earthquake insurance. In response, California modified its requirements to allow sellers of homeowners' insurance to instead offer reduced, supplementary earthquake coverage through the state-sponsored California Earthquake Authority. By lessening insurers' exposure to catastrophes, California was able to induce more insurers to offer homeowners' and earthquake policies, but only after the Northridge crisis had passed.

States also succeeded in delaying rate increases immediately following those catastrophic events. According to the Florida Department of Insurance, premiums for Florida's homeowners doubled, on average, between 1992 and 1999. But those higher premiums do not yet fully reflect the increase in expected losses from future natural disasters. (Because of rate suppression, state-sponsored insurers still assume much of the risk in Florida and California.) In fact, rates in Florida remain below those needed to attract enough insurers to reduce the size of the state's Windstorm Underwriting Association, an insurer of last resort that has taken over most of the market for hurricane insurance in high-risk areas.[23]

In the short run, state-sponsored funds may make coverage more available and affordable. But to the extent that those funds keep prices below costs and crowd out private suppliers, insurance may be harder to obtain and more expensive in the long run. The rate structures also force property owners in lower-risk areas to subsidize the cost of hurricanes and earthquakes in higher-risk areas. Those cross-subsidies reduce incentives for mitigation, leading to higher losses over time. Large future losses can in turn expose taxpayers to risk if the state government chooses to bail out insolvent insurance pools.

Whereas too much regulation can potentially harm an insurance market, a relative lack of regulation can sometimes prove beneficial. Such an outcome can be seen in California's large and well-functioning private market for commercial earthquake insurance. Relative to the state's market for residential earthquake insurance, its much-larger market for commercial earthquake insurance is lightly regulated, especially with respect to price. Insurers are also free to choose their customers geographically and thus avoid excessive concentration of risk. Even global reinsurance companies are active in providing earthquake coverage for commercial buildings in California.

## Federal Assistance After Disasters

Through emergency supplemental appropriations, the Congress often provides extensive financial assistance to areas with uninsured losses after a disaster. Following Hurricanes Andrew and Iniki in 1992, the federal government spent more than $4.4 billion on the affected areas. Similarly, the region struck by the Northridge earthquake received $9.5 billion in aid (*see Table 5*).

Under the Budget Enforcement Act of 1990 (BEA), new spending designated by the President and the Congress as an emergency is exempt from the BEA's limits, or caps, on discretionary spending and from pay-as-you-go requirements for mandatory spending and receipts. (The caps on discretionary spending expire in 2003.) That budgetary treatment provides a sort of safety valve for emergencies; once the spending has been enacted, the discretionary spending caps are adjusted upward by the same amount.[24]

Federal assistance is disbursed through various departments and agencies. The Federal Emergency Management Agency (FEMA) provides public assistance grants to state

*1998 Florida Insurance FACT Book* (Tallahassee, Fla.: Florida Insurance Council, April 1998), p. 63. The moratorium expired on July 1, 2001, but it may be renewed in 2002.

23. After Hurricane Andrew, Florida allowed insurers to substitute the state's windstorm insurance for their own hurricane coverage in certain high-risk regions. The Windstorm Underwriting Association currently has almost 410,000 policies, with an exposure exceeding $98 billion. Like all residential insurers in Florida, the association purchases reinsurance from the state's catastrophe fund. Arbitrators, who hear challenges to rate decisions by the Department of Insurance, recently approved a 96 percent rate hike for the Florida Windstorm Association. However, that rate hike was challenged and remains before the Florida courts.

24. See Congressional Budget Office, *Emergency Spending Under the Budget Enforcement Act*, CBO Memorandum (December 1998); and Congressional Budget Office, *The Budget and Economic Outlook: Fiscal Years 2001-2010* (January 2000), Box 1-1, p. 12.

**Table 5.**

## Federal Financial Assistance After Selected Disasters

(In millions of dollars)

| | Hurricane Andrew[a] | Northridge Earthquake |
|---|---|---|
| Department of Agriculture | 1,057[b] | n.a. |
| Department of Commerce | 101 | n.a. |
| Department of Defense[c] | 839[b] | n.a. |
| Department of Education | 102 | 245 |
| Department of Health and Human Services | 106 | n.a. |
| Department of Housing and Urban Development | 208[b] | 825[d] |
| Department of Transportation[e] | 90[b] | 950[d] |
| Subtotal | 2,503 | 2,020 |
| Federal Emergency Management Agency | | |
| Public Assistance Grants | 823 | 4,507 |
| Temporary Housing | 128 | 1,199 |
| Mitigation Grants | 23 | 738 |
| Individual and Family Grants | 177 | 165 |
| Other | 83 | 65 |
| Subtotal | 1,234 | 6,674 |
| National Flood Insurance Program | 168 | n.a. |
| Small Business Administration[f] | | |
| Loans to Individuals | 106 | 429 |
| Loans to Businesses | 71 | 270 |
| Salaries and Administrative Costs | 100 | 55 |
| Subtotal | 277 | 754 |
| Other Departments and Agencies | 228[b] | 67[d] |
| **Total** | **4,410** | **9,515** |

Source: Congressional Budget Office based on information from the Federal Emergency Management Agency (FEMA), the Small Business Administration (SBA), and the Congressional Research Service.

Note: n.a. = not applicable.

a. Totals for Hurricane Andrew also include some spending for Hurricane Iniki and Typhoon Omar. The amounts listed for FEMA and SBA cover only Hurricane Andrew and exclude administrative costs.

b. Amounts appropriated in the 1992 Dire Emergency Supplemental Appropriations Bill Covering Hurricanes Andrew and Iniki and Typhoon Omar. Excludes amounts totaling $991 million for which a specific budget request was necessary.

c. Includes $197 million specifically designated for repairing and maintaining U.S. air force and navy bases in Guam, which was hit by Typhoon Omar. Homestead Air Force Base in Florida was hit by Hurricane Andrew.

d. Amounts directly appropriated in the 1994 Dire Emergency Supplemental Appropriations Bill Covering Northridge Earthquake. Excludes $900 million in spending for which a subsequent official budget request was necessary.

e. Includes amounts from the federal Highway Trust Fund.

f. CBO estimated the cost of SBA's credit programs using reestimates of the long-term expected costs of those loans. The reestimates are considerably less than the original estimates.

and local governments to repair damaged infrastructure, such as bridges, highways, and public buildings. The Small Business Administration helps individuals pay for temporary housing and offers both direct and guaranteed loans to businesses. The federal crop insurance programs convey additional assistance to farmers.

The $40 billion in emergency appropriations passed by the Congress in response to the September 11 attacks included about $10.2 billion for New York City.[25] An estimated $4.3 billion of that amount (distributed through FEMA) will be used to reclaim the World Trade Center site and to rebuild transit systems and government buildings. The remainder comprises loans and assistance to businesses affected by the attacks, unemployment assistance for displaced workers, and health assistance and monitoring. Other laws provide about $5.8 billion in compensation for victims and their families. Specifically, the Air Transportation Safety and System Stabilization Act (Public Law 107-42) provides $5.4 billion in compensation; additional assistance comes from the USA PATRIOT Act (P.L. 107-56), the Aviation and Transportation Security Act (P.L. 107-71), and the Victims of Terrorism Tax Relief Act (P.L. 107-134). The President has promised more financial assistance to New York City.

Federal disaster assistance undoubtedly reduces financial hardships, but it may also discourage individuals and state and local governments from purchasing adequate insurance against future losses. In effect, it subsidizes development in disaster-prone areas (ones for which private insurers might be reluctant to provide coverage), and it weakens people's incentives to take actions that would reduce the cost of future natural disasters. Critics also contend that the "emergency" designation constitutes a large spending loophole that weakens overall budgetary discipline.

25. Congressional Budget Office, *The Budget and Economic Outlook: Fiscal Years 2003-2012* (January 2002), pp. 111-120.

# 3

# Proposals for Federal Reinsurance

Two groups of federal policies have been proposed to increase the supply of property and casualty insurance. The first group—which includes H.R. 21 (Homeowners' Insurance Availability Act of 1999) and H.R. 230 (Natural Disaster Protection and Insurance Act of 1997) —was developed after Hurricane Andrew and the Northridge earthquake. Those policies aim to address the perceived shortage of new insurance against losses from natural disasters. The second group of policies, which comprises H.R. 3210 (the Terrorism Risk Protection Act) and S. 2600 (the Terrorism Risk Insurance Act of 2002) responds to the market disruption following September 11. Both groups of proposals could be considered alternative federal responses, regardless of the source of the catastrophic loss.

The principal distinction between the groups of policies centers on the role of the federal government. Proposals to increase the supply of insurance for natural disasters would have the government serve as a commercial reinsurer, in effect. The U.S. Treasury would sell one-year reinsurance contracts to primary insurers and reinsurers, as well as to state-sponsored plans, through competitive auctions. Bids would start at a set price. The policies' intent is to offer reinsurance when it is in short supply, at market prices that are expected to cover the government's costs.

By contrast, proposals developed since September 11 would have the government take on risks without full compensation from insurers or policyholders. Under one variation, the government would pay for losses above a certain level, without receiving any compensation. Under another, the government would require that some portion of any assistance be repaid by all participating insurers and policyholders over a number of years. In the case of terrorism insurance, the policies' intent is to have taxpayers subsidize insurers and policyholders by bearing most of the cost of terrorism-related losses.

## Arguments For and Against Federal Intervention

Arguments for a federal role in the insurance market generally fall into two categories. The first category of arguments emphasizes the high cost and limited supply of private capital.[1] According to proponents, a shortage of risk-bearing capital leads to an inadequate supply of insurance, which keeps prices steep relative to projected losses for low-probability, high-loss events. One analyst estimated that catastrophe reinsurance premiums were more than seven times the expected loss in 1994, on average, although that multiple had dropped to between three and four by 1998.[2] Proponents also contend that the

1. For example, see David M. Cutler and Richard Zeckhauser, "Reinsurance for Catastrophes and Cataclysms," in Kenneth A. Froot, ed., *The Financing of Catastrophe Risk* (Chicago: University of Chicago Press, 1999), pp. 233-269.

2. Premiums for the highest layers of coverage (the lowest probability layers) were between 20 and 30 times expected losses in 1994, according to estimates. See Kenneth A. Froot, *The Market for Catastrophe Risk: A Clinical Examination,* Working Paper No. 7286 (Cambridge, Mass.: National Bureau of Economic Research, February 2001), p. 7 and Figures 3 and 4. However, research emphasizes the imprecision of the estimates of actuarial losses for the least likely events. See James F. Moore, *Tail Estimation and Catastrophe Security Pricing: Can We Tell What Target We Hit If We Are Shoot-*

federal government, with its vast capacity to tax and borrow, has an advantage over private insurers in bearing catastrophe risk because it does not need to hold explicit capital to pay off claims and avoid bankruptcy.[3] To free insurers from the costly burden of holding huge amounts of capital, proponents suggest that the government act as a residual provider of reinsurance for so-called mega-catastrophes (disasters costing in the $100 billion range). The government could set premiums below those charged by private insurers, thus lowering the cost of insurance while protecting taxpayers from losses.

The second category of arguments emphasizes that the biggest barrier to an adequate supply of insurance, especially immediately after a catastrophe, is insurers' heightened uncertainty about the frequency and size of future losses. After Hurricane Andrew, the Northridge earthquake, and the World Trade Center attacks, insurers were not certain that they could assess the risks they were being asked to assume.[4] Without such knowledge, they were unwilling to commit capital by issuing coverage.[5]

In time, insurers are usually able to recalibrate their estimates and reenter the market. (Some insurers have already begun to do so on a limited basis in the aftermath of the terrorist attacks.)[6] Thus, proponents contend that the government needs to intervene to supply insurance while insurers reassess risk after a disaster, but they argue for a temporary federal role.

Many counterarguments have been made against the claim that federal intervention would improve the performance of insurance markets.[7] For example, some analysts dispute the claim that insurers' unwillingness to provide coverage after a disaster is a market failure. Instead, they argue that the absence or high price of insurance implies that the proposed activity is too risky. (Prices often seem higher to buyers of insurance, who just see the rate hikes, than to capital providers, who understand the relationship of price to risk.) Defenders of the market also assert that capital flows to investments for which the expected rates of return, after adjusting for risk, are highest. Thus, shortages of capital simply reflect prices (and expected rates of return) that are too low. And many analysts are skeptical of the claim that the federal government could establish a temporary reinsurance program, contending that it would become a permanent barrier, subsidized by taxpayers, to the provision of insurance by private firms.

In addition, opponents of intervention argue that the federal government does not have a cost advantage in bearing risk and reinsuring catastrophes.[8] Private insur-

---

*ing in the Dark?* Working Paper No. 99-14 (Philadelphia: Wharton Financial Institutions Center, 1999), available at http://fic.wharton.upenn.edu/fic/.

3. Statement of Lawrence H. Summers, Deputy Secretary, Department of the Treasury, before the House Banking and Financial Services Committee, April 23, 1998.

4. For an analysis of the impact on risk assessment of the September 11 events, see the statement of Alan Greenspan, Chairman, Federal Reserve Board of Governors, before the Joint Economic Committee, U.S. Congress, October 17, 2001. Also see the statement of Kenneth A. Froot, Professor of Finance, Harvard Business School, before the Senate Committee on Banking, Housing, and Urban Affairs, October 24, 2001.

5. After the September 11 attacks, actions by rating agencies reinforced that uncertainty. At least one firm said it would probably downgrade the credit ratings of insurers who continued to cover terrorism risks. Downgrades raise an insurer's cost of capital. See Standard & Poor's, "Maintenance of Insurance Ratings Depends on Mitigating Terrorism Risk," *Ratings Direct*, October 19, 2001.

6. Insurers are thinking about how to model terrorism risk. See John A. Major, "Advanced Techniques for Modeling Terrorism Risk" (February 2002), available at www.guycarp.com/pdf/major_terrorism.pdf; and Gordon Woo, "Quantifying Insurance Terrorism Risk" (paper presented at the National Bureau of Economic Research Conference on Insurance, Cambridge, Mass., February 1, 2002).

7. For example, see Shadow Financial Regulatory Committee, "Proposed Federal Catastrophe Reinsurance," Statement No. 158 (December 13, 1999), available at www.aei.org/shdw/shdw.158.htm; and Shadow Insurance Regulation Committee, "The Efficient Management of Catastrophic Risk" (Washington, D.C.: Shadow Insurance Regulation Committee, March 1, 1999).

8. Scott E. Harrington, "Rethinking Disaster Policy," *Regulation*, vol. 23, no. 1 (2000), pp. 40-46; George Priest, "The Government, the Market, and the Problem of Catastrophic Loss," *Journal of Risk and Uncertainty*, vol. 12, no. 2/3 (1996), pp. 219-237; and R. Anton Braun, Richard M. Todd, and Neil Wallace, *The Role of Damage-Contingent Contracts in Allocating the Risks of Natural Catastrophes*, Working Paper No. 586D (Minneapolis, Minn.: Federal Reserve Bank of Minneapolis, April 1998).

ance firms are agents for investors who have a relatively high tolerance for risk. In contrast, federal reinsurance would place risks on taxpayers indiscriminately, regardless of their willingness or ability to bear them. Federal reinsurance may appear to be less costly, but that is only because it does not include the costs of compensating taxpayers for the burden of supplying capital.

Because terrorism risks differ from the risks of natural disasters, some analysts believe that subsidizing terrorism insurance is the public's responsibility. The federal government has the leading role in protecting the nation from terrorism, and the September 11 attacks blurred the line between homeland security and national defense in the budget.[9] Proponents of federal intervention assert that losses from terrorist attacks are part of the cost of providing homeland security because targets may be specifically chosen according to the patriotic or iconic value that the public attaches to them.

## How Would Federal Reinsurance of Risks from Natural Disasters Work?

Advocates of the view that the market cannot provide sufficient capital to protect property against natural disasters would establish a federal program to sell reinsurance to state-sponsored and private insurers and reinsurers.[10] (See Appendix C for an analysis of government-backed programs in Japan and New Zealand that provide insurance against earthquakes.) Early versions of proposals that have the government providing reinsurance covered both residential and commercial properties; most later versions apply only to residential ones. The proposals generally contain provisions intended to control federal spending and avoid crowding out the private sector.[11] Those provisions would:

- Require that the reinsurance contracts be auctioned, using minimum prices based on actuarially expected losses plus an additional amount for risk load and administrative costs;[12]
- Define a minimum level of insured losses (a "retention level") that must be sustained in each state or region before the federal government would make payments;
- Limit the federal government's maximum liability; and
- Set a date on which the program would expire.

Most proposals also demand that states increase their efforts to reduce risk through mitigation; some proposals go further, requiring that a small portion of the proceeds from the sale of insurance go into a federal mitigation trust fund, which would make money available to eligible states for those efforts. For example, Florida could subsidize the purchase of storm shutters to protect windows from high winds and windblown debris, and California could subsidize the cost of bolting or bracing walls to a home's foundation to make the structure more resistant to earthquake damage.

### Auctioning Contracts with a Minimum Bid Price

The federal government would auction reinsurance contracts to eligible private insurers and reinsurers. The proposed contracts, which would be subject to a minimum bid or reservation price, would pay a fixed amount per

9. See the exchange between Alan Greenspan, Chairman, Federal Reserve Board of Governors, and Representative Carolyn Maloney in the transcript of hearings before the House Financial Services Committee, February 27, 2002.

10. Proposals include H.R. 21 and S. 1361 in the 106th Congress; H.R. 219, H.R. 230, H.R. 579, and S. 1043 in the 105th Congress; and H.R. 1856 in the 104th Congress. None of those proposals is currently being considered.

11. For an analysis of the proposals, see Christopher M. Lewis and Kevin Murdock, "Alternative Means of Redistributing Catastrophic Risk in a National Risk-Management System," in Kenneth A. Froot, ed., *The Financing of Catastrophe Risk* (Chicago: University of Chicago Press, 1999), pp. 51-85; and Lewis and Murdock, "The Role of Government Contracts in Discretionary Reinsurance Markets for Natural Disasters," *Journal of Risk and Insurance*, vol. 63, no. 4 (1996), pp. 567-597.

12. In the private sector, the risk-load factor is an amount added to the risk-based price to compensate the insurer for the variability of losses around the long-term average in any given year and for the uncertainty surrounding available estimates of the average itself.

$1 billion of losses stemming from a disaster.[13] The contracts would be similar to the catastrophe options traded on the Chicago Board of Trade. They would cover specific "layers" of loss and be divisible and tradable. Proposals typically require that separate auctions cover different geographic regions and that Florida and California each constitute at least one region.

Under most of the proposals, a national commission of actuaries would determine the expected losses for each region. Proposals typically direct that prices for the reinsurance contracts include a risk load—often 100 percent to 200 percent of the expected loss—mainly to account for the variability and uncertainty of claims and the government's cost of funds. (Higher risk loads could be added to contract prices if the program's administrator concluded that they were warranted.) All other things being equal, the risk load for a catastrophic loss that happens once every 250 years, on average, should be higher than the risk load for a loss that occurs once every 100 years.[14]

The contracts for state-sponsored programs would not be auctioned under some of the proposals; rather, prices would be set by the Treasury. In general, the proposals require that prices be set at least as high as the combined total of expected losses and administrative costs.

### Setting Retention Levels

Because a federal reinsurance program is designed to cover only the highest levels of losses, insurers would have to sustain large losses before the federal contracts paid out money. Under most proposals, neither Hurricane Andrew nor the Northridge earthquake would have triggered payments. The retention levels and contract payouts are tied to overall, or aggregate, losses rather than to losses for individual insurance companies. The use of aggregate loss triggers has several advantages: it vastly simplifies the program's administration, provides incentives for insurers to maintain high standards for underwriting and claims adjustment, and facilitates trading of the securities in the secondary market.

But using aggregate loss triggers also means that any particular insurer's losses may not match the contract's payout. That mismatch—or basis risk—reduces the value of the contracts as reinsurance because insurers could be left underexposed or overexposed to risk. Indeed, catastrophe options with similar features failed to attract much interest on the Chicago Board of Trade (see Appendix B). Under some early reinsurance proposals, all holders of the contracts would receive payments from the federal government regardless of whether they had actually incurred losses. However, under more recent proposals, only insurers with losses would receive compensation.

Federal reinsurance proposals use three types of triggers to describe the magnitude of catastrophe necessary to activate contract payments. Under an aggregate loss trigger, payments would occur if total insured losses from one or more catastrophes exceeded a set level, such as $25 billion a year under a national contract. (One problem with that type of "excess of loss" reinsurance is that initial estimates of losses from earthquakes may be wildly inaccurate. For example, initial estimates from the Northridge earthquake were just $2.5 billion; ultimately, the insured losses were $15 billion. Thus, payouts under a federal program of that type might be slow to materialize.) Under an event frequency trigger, contract payments would result following a catastrophe whose expected loss frequency exceeded a certain threshold, such as every 100 years. Event frequency is often determined by regional insured losses. For instance, a loss with a 100-year frequency might be identified as one that caused property damage of $6 billion in an area less prone to disasters but $20 billion in a more high-risk area. Finally, in the case of contracts sold to state-sponsored insurers, the level at which payments were triggered could be set by the purchaser's total payment capacity. For example, contracts sold to a state-sponsored insurer with a total payment capacity of $5 billion would

13. Under some proposals, such as S. 1361, the contracts would be issued by nonprofit corporations with special ties to the government, including the ability to borrow from the Treasury. That structure increases the risk that the entity could use public benefits for its private gain. See the statement of Stuart Eizenstat, Deputy Secretary, Department of the Treasury, before the Senate Committee on Commerce, Science, and Transportation, April 13, 2000.

14. For an analysis of the factors determining risk loads, see the statement of Paul G.J. O'Connell, Chair, Catastrophe Working Group, American Academy of Actuaries, before the House Banking and Financial Services Committee, April 23, 1998.

pay out if a catastrophe caused the insurer to incur claims exceeding that amount.

### Capping the Contingent Liability

To protect taxpayers, some proposals include provisions to limit the government's annual payment obligation. Such limits are sometimes defined by a maximum level of total payments—for instance, $25 billion a year—which would constrain the number of contracts sold. Alternatively, a proposal might direct the program's administrator to restrict the amount of coverage sold each year so that total annual contract payments were "unlikely" to exceed a specified level, such as $25 billion. The imprecision of those caps reflects the uncertainty surrounding the number of regions and state programs that might be covered, the minimum level of insured losses that must be sustained, and the layers of losses that would be covered, as well as the possibility of more than one catastrophe in a year. Many analysts believe that it would not be politically feasible to pay less than the full amount on contracts sold and that any cap would be moot in the event of multiple disasters in the same year.[15] Furthermore, using "unlikely" as a standard would set no practical limit on contract volume because catastrophes are, by definition, unlikely.

### Putting It All Together: An Example

Suppose the government was to offer reinsurance on a national contract covering the layer of losses between $25 billion and $50 billion from a single disaster. The private sector and state-sponsored plans would pay for losses under $25 billion (the deductible) and over $50 billion (the upper limit).[16] The government could auction 1,000 contracts that would each pay insurers $1 million for every billion dollars of industry losses in that layer.

Consider scenarios for three different disasters, causing insured losses of $23 billion, $33 billion, and $60 billion, respectively. For the $23 billion loss, no money would be paid out because the loss would be below the deductible. For the $33 billion loss, each contract would pay out $8 million, so the government's payment on 1,000 contracts would total $8 billion. For the $60 billion loss, each contract would pay out $25 million (the maximum amount), capping the government's contingent liability at $25 billion.

Setting a minimum contract price could prove tricky because of the difficulty of determining expected losses. Under a national excess-of-loss contract for a layer of losses between $25 billion and $50 billion, the expected loss component that would be included in bid prices could range from $350,000 to $1,750,000 annually for $25 million in coverage.[17] The width of that range—the high end is five times as large as the low end—indicates the uncertainty surrounding estimates of actuarial losses.

The design of a federal reinsurance contract will in part determine its viability. For example, a national contract's ability to attract sufficient demand depends partly on rules governing trading. Suppose that the event triggering a payout is an earthquake in California. A firm with limited or no exposure in the California market could emerge as a "winner" by selling its contract to a firm with heavy losses from the earthquake. (The federal government could restrict trades to prohibit those kinds of windfalls, but that action would reduce demand for the contracts.) In private markets, most reinsurance transactions are negotiated on an individual basis to avoid a mismatch between insurers' losses and reinsurance payouts.

15. See the statement of Stuart Eizenstat, Deputy Secretary, Department of the Treasury, before the House Banking and Financial Services Committee, July 30, 1999.

16. Under some proposals, the government would provide only 50 percent coverage rather than 100 percent for a specified layer of losses. That kind of coinsurance has two advantages. First, it provides a stronger incentive for firms to be more vigilant against fraud and inflated claims. Second, it allows more opportunity for private reinsurers to issue policies.

17. That range is for the best-fitting estimates. See J. David Cummins, Christopher M. Lewis, and Richard D. Phillips, "Pricing Excess-of-Loss Reinsurance Contracts Against Catastrophic Risk," in Kenneth A. Froot, ed., *The Financing of Catastrophe Risk* (Chicago: University of Chicago Press, 1999), pp. 93-141, especially Table 3.8, p. 132.

## Cost to the Federal Government of Reinsuring Risks from Natural Disasters

The Congressional Budget Office (CBO) concludes that all of the current reinsurance proposals would be costly to taxpayers, although the proposals' specific budgetary impact is uncertain. In particular, a proposed 100 percent or 200 percent risk load would be unlikely to provide an adequate cushion against losses on an expected-value basis (over a long time horizon).[18] By comparison, risk loads in private transactions are typically four to six times (that is, 400 percent to 600 percent) expected losses.

CBO's conclusion rests on two factors: the uncertainty surrounding the actuarial estimates of catastrophe risk that would be used to set minimum contract prices, and the federal government's weak incentives to overcome that uncertainty and set prices at a budget-neutral level. For example, the actuarially expected losses for the $25 billion to $50 billion layer of losses on a national contract could vary greatly depending on which type of probability distribution of losses was chosen to fit the meager available data.[19] Improvements in building techniques over time would also add to the uncertainty because those changes would make the historical record of losses less reliable as a predictor. In addition, the risks themselves might be shifting. Meteorologists believe that changes in ocean cycles and atmospheric conditions may be increasing the frequency of stronger hurricanes.[20]

Given the irreducible uncertainty of the estimates, property owners and insurers—bringing consumer and political pressure to bear—could probably generate a strong voice for low contract prices. (Political pressure could be especially influential if prices were set administratively for the state-sponsored programs.) Another option, auction bidding, might drive contract prices toward the break-even value, but CBO is not certain that demand for the contracts would be sufficient to ensure vigorous bidding. The recent halt in trading of catastrophe options on the Chicago Board of Trade adds to that uncertainty.[21]

Over a 10-year period, a federal reinsurance program with annual contracts covering losses that occurred once every 100 years, on average, for six different regions would have 60 chances to encounter a catastrophic event; the probability of at least one such event occurring would be 45 percent. (Some proposals allow the Secretary of the Treasury to lower the loss threshold for state-sponsored programs.) Furthermore, the probability of such a 100-year loss may be more (or less) than 1 percent in any given year, either because the historical data underlying the estimates of the frequency of events are inadequate or because the timing of such events is affected by meteorological cycles.

If a catastrophic earthquake or hurricane occurred, the federal government would probably be left paying for some of the losses. It would be difficult for the government to raise premiums by enough to recover past losses and repay the money it borrowed from the Treasury over the time horizons in the proposals, which vary from 10 years to 20 years. Indeed, under competitive conditions, it would be nearly impossible for the government to recoup past underwriting losses that were based on mispriced premiums. Firms that did not participate in the program would be able to underprice participating firms and capture most of the market, unless the costs of past losses were also imposed on new entrants. Moreover,

---

18. See Congressional Budget Office, *Cost Estimate for H.R. 21, Homeowners' Insurance Availability Act of 1999* (February 9, 2000); *Cost Estimate for H.R. 219, Homeowners' Insurance Availability Act of 1998* (September 23, 1998); and *Cost Estimate for H.R. 230, Natural Disaster Protection and Insurance Act of 1997* (October 8, 1997).

19. Researchers tested four different loss distributions. The estimates of expected loss ranged from $177 million (using a lognormal distribution and adjusting past losses for population growth) to $4,635 million (using a Pareto distribution and simulating losses). See Cummins, Lewis, and Phillips, "Pricing Excess-of-Loss Reinsurance Contracts Against Catastrophic Risk," pp. 93-141, especially Table 3.4 on p. 123 and Table 3.6 on p. 127.

20. See Andrew C. Revkin, "Coasts and Islands Facing Era of Strong Hurricanes," *New York Times*, July 20, 2001, p. A1; and Stanley B. Goldenberg and others, "The Recent Increase in Atlantic Hurricane Activity: Causes and Implications," *Science*, vol. 293 (July 20, 2001), pp. 474-479.

21. Swiss Re, "Capital Market Innovation in the Insurance Industry," *Sigma*, no. 3 (2001), pp. 21-23, available at www.swissre.com in the *Sigma* archive.

firms would have an incentive to reduce their purchases of reinsurance from the government to hold down prices.

Some of the costs to the federal government could be offset by possible savings in payments for disaster assistance, particularly assistance provided by the Small Business Administration. However, several of the largest disaster-assistance programs benefit individuals or organizations that would not be affected by the reinsurance contracts (*see Table 5* on page 16). For instance, most spending from the Federal Emergency Management Agency's disaster-relief fund goes to state and local governments, which do not carry traditional insurance, to aid in rebuilding public infrastructure. Federal costs for disaster assistance might also be reduced by provisions in the reinsurance proposals that increase spending on preventive measures, but the size of any such savings would be speculative.

## How Would Federal Reinsurance of Risks from Terrorism Work?

In countries where terrorist attacks have occurred, such as Great Britain, Israel, and Spain, government-backed programs provide insurance coverage.[22] Under the proposals now before the Congress, the federal government would bear most of the cost of terrorism-related losses for the next one to three years. Neither the House bill (the Terrorism Risk Protection Act, or H.R. 3210) nor the Senate bill (the Terrorism Risk Insurance Act of 2002, or S. 2600) envisions a permanent program, but one could result, especially if the threat of terrorism remains high. The two bills differ in their deductibles, the share of risk assumed by the government, the caps on punitive damages, and the amounts of repayment.[23] Both bills cover claims for business interruption and workers' compensation and limit the government's annual liability to $100 billion. If losses exceeded that amount, the Secretary of the Treasury could ask the Congress for guidance on how losses over the limit should be paid.

The Terrorism Risk Protection Act, which the House passed in November 2001, requires the federal government to coinsure, or bear, 90 percent of all losses after the insurance industry sustains $1 billion in total losses in a year. The overall level of loss at which the government would intervene could be as low as $100 million if the losses for any single insurer exceeded 10 percent of its capital surplus and 10 percent of its net premiums. Each insurer would also have a deductible of $5 million. Insurers and policyholders as a group would partially repay the government for its financial assistance. For losses of less than $20 billion, property and casualty insurers would pay assessments of up to $15 billion to the government—up to $5 billion in the first year and up to $10 billion in subsequent years (annual assessments cannot exceed 3 percent of an insurer's premiums).[24] Commercial property and casualty policyholders would repay losses of between $20 billion and $100 billion that were sustained by the government through surcharges of up to 3 percent on their premiums. The government would subsidize the repayment by not charging interest on the funds it provided to insurers. Of course, whether the assessments are levied on policyholders or insurers, policyholders ultimately bear the burden of repayment. However, assessments that were levied directly on insurers would affect their balance sheets.

The Terrorism Risk Insurance Act of 2002, which the Senate passed in June 2002, departs from the House bill in two major ways. First, the federal government would not be repaid for its losses. Second, in the event of a catastrophe during the program's first year, the industry would

22. Statement of Thomas J. McCool, Managing Director, Financial Markets and Community Investment, General Accounting Office, before the Senate Committee on Banking, Housing, and Urban Affairs, published as General Accounting Office, *Terrorism Insurance: Alternative Programs for Protecting Insurance Consumers*, GAO-02-199T (October 24, 2001).

23. For a detailed comparison of the various bills before the Congress, see Rawle O. King, *Terrorism Risk Insurance: A Summary of Legislative Proposals*, CRS Report for Congress RL31209 (Congressional Research Service, December 7, 2001).

24. Under the insurance industry's accounting rules, an insurer would have to recognize a liability at the time of the loss for the full future value of the payments due. If the liability was large enough, a participating company could become insolvent, even if it suffered no losses. Consequently, some analysts doubt whether H.R. 3210 would encourage insurers to offer substantially more terrorism coverage than they offer now. See Standard & Poor's, "Terrorism Coverage Remains in Doubt," *Ratings Direct*, April 15, 2002.

have to absorb the first $10 billion (rather than $1 billion) in losses. After the industry had paid $10 billion, the government would pay 90 percent of the remaining losses, up to $100 billion. If the Treasury extended the program for a second year, insurers would have to cover the first $15 billion in losses.

Under the provisions of the Senate bill, federal cost sharing could occur for losses of less than $10 billion if the solvency of individual firms was at risk. A firm's deductible in the first year of the program would be its market share multiplied by the $10 billion industry deductible. Thus, a firm with 5 percent of the market would face a $500 million deductible, after which the federal government would pay 80 percent of the losses up to the $10 billion industrywide trigger.

## Cost to the Federal Government of Reinsuring Risks from Terrorism

CBO cannot predict how much insured damage terrorist attacks could cause in any specific year. Thus, the agency's estimate of the cost of financial assistance provided under H.R. 3210 reflects how much the government might be expected to pay insurers on average. That average weighs the probabilities of various outcomes, from damages of zero up to very large amounts, resulting from possible future terrorist attacks. On such an expected-value basis, CBO estimates that H.R. 3210 would increase direct spending by $8.5 billion and government receipts by $5.3 billion over the 2002-2011 period.[25] That estimate is based on premiums collected for terrorism insurance in Great Britain and insurance practices in the United States. (See Appendix C for information on Britain's terrorism reinsurance program.) It assumes that financial assistance to property and casualty insurers would be paid over several years because of disputes by property and business owners about the value of covered losses.

25. Congressional Budget Office, *Cost Estimate for H.R. 3210, Terrorism Risk Protection Act* (November 19, 2001), as ordered reported by the House Committee on Ways and Means on November 16, 2001. The bill passed by the House differed slightly. CBO did not prepare a cost estimate for S. 2600.

The federal government would need to charge insurers about $3 billion annually for coverage over the 2002-2004 period, CBO estimates, to receive full compensation for the risk it would assume under the Terrorism Risk Protection Act. Although that estimate reflects CBO's best judgment based on available information, costs are a function of future terrorist attacks, which are inherently unpredictable. As such, actual costs could cover an extremely broad range.

## Advantages and Disadvantages of the Federal Reinsurance Contracts

Reinsurance proposals seek to expand the capacity of private insurance markets to handle catastrophes without shifting costs to taxpayers or preempting private-sector initiatives. Ancillary goals include mitigating losses before disasters strike, discouraging insurance-related discrimination, and encouraging states to take steps to combat price-gouging after disasters. Federal excess-of-loss reinsurance contracts could expand insurance coverage and reduce some federal spending on disaster assistance. They could also back up states' disaster plans. In the short run, federal reinsurance of terrorism risks may be the only way to ensure that significant amounts of traditional insurance are available for buildings that are landmarks or for critical infrastructure projects.[26] If the reinsurance contracts were underpriced, however, they could reduce incentives for private mitigation, preempt the development of innovative insurance products, and shift costs to taxpayers.

### Advantages

Auctioning excess-of-loss contracts would probably lower the cost of reinsurance, which would in turn increase the availability and decrease the price of catastrophe insurance. For that reason, primary insurance companies gen-

26. Rating agencies have stated that exposure to terrorism losses could result in ratings downgrades in the absence of a federal insurer of last resort. See Steven J. Dreyer, Mark Puccia, and Donald Watson, "Maintenance of Insurance Ratings Depends on Mitigating Terrorism Risks," Standard & Poor's, *Ratings Direct*, October 19, 2001; and Keith M. Buckley, "Fitch Comments on Possible Government Terrorism Solution," Fitch Special Report (November 12, 2001).

erally support a federal program.[27] The reinsurance contracts would confer benefits both before and after a disaster. They would create a safety net for private insurers and reinsurers before disaster struck by absorbing the risk of losses from megacatastrophes, which otherwise could threaten firms with insolvency.[28] By covering losses and providing liquidity after a disaster, a federal program would make insurers more willing to continue issuing coverage in disaster-prone areas.

By relying on auctions rather than set prices to determine the cost of the contracts, the federal government recognizes that it is probably at an informational disadvantage relative to private insurers in pricing catastrophe risk. Auctions offer the government a way of potentially neutralizing its disadvantage because they can compel insurers to reveal their assessment of risk in their bids. (Insurers whose bids understated their assessment of risk might not win any of the reinsurance contracts.)[29]

Competition among insurers would force them to pass the savings from a federal program on to policyholders, many of whom would see their premiums drop. Policyholders might not realize any savings in states in which regulators suppress rates substantially below costs, but the quality of the insurance product could improve. Similarly, federal reinsurance for disasters could enhance the credibility of states' insurance programs and reduce their reliance on postevent assessments.

27. See the statements of Robert W. Pike, Senior Vice President, Allstate Insurance Company, before the House Committee on Banking and Financial Services, April 23, 1998, and July 30, 1999. Reinsurers also support a federal role provided that the retention levels are set high enough to avoid crowding them out. See the statement of Franklin W. Nutter, President, Reinsurance Association of America, before the House Committee on Banking and Financial Services, April 23, 1998.

28. See Rade T. Musulin, "Would a Federal Role in Disaster Protection Be a Catastrophe?" *Contingencies*, American Academy of Actuaries (November/December 2000), pp. 28-33.

29. The government is more likely to extract private information on the value of the contracts if it sets supply below expected demand at the reservation price. See Lewis and Murdock, "The Role of Government Contracts in Discretionary Reinsurance Markets for Natural Disasters."

By encouraging more insurers to offer greater coverage, federal reinsurance might also reduce people's need for federal assistance after disasters. In particular, individuals' need for loans through the Small Business Administration (which cost taxpayers more than $500 million in total after Hurricane Andrew and the Northridge earthquake) and their need for temporary housing assistance through FEMA (which cost more than $1.3 billion for the two natural disasters) could decline (*see Table 5* on page 16).

A federal reinsurance program for terrorism risks could provide two additional benefits. First, it could reduce the short-term economic effects of the current reduction in the supply of insurance. Second, it would spread the costs of the threat of terrorism across the country, rather than leave those costs to be borne by the owners and users of particular facilities. People who view losses from terrorist attacks as part of the cost of providing homeland security would view such cost sharing as an advantage.

**Disadvantages**

The federal government's experience with other insurance programs suggests that it has difficulty managing risk efficiently. In part, that is because the government has less of an incentive than private firms do to ensure that premiums cover costs and to control moral hazard (policyholders' incentive to change their behavior in ways that will increase losses from insured events) and adverse selection (the purchase of insurance disproportionately by people with the highest risks).

One way in which that difficulty manifests itself is in oversimplified pricing; federal programs seldom tailor premiums to risks. For example, the federal flood insurance program charges premiums that are below expected costs for some properties and covers repeated losses for the same properties.[30] Another example is evident in the

30. See the statement of Stanley J. Czerwinski, Associate Director, Housing and Community Development Issues, Resources, Community, and Economic Development Division, General Accounting Office, before the Subcommittee on Housing and Community Opportunity of the House Committee on Banking and Financial Services, published as General Accounting Office, *Flood Insurance: Information on Financial Aspects of the National Flood Insurance Program*, GAOT-RCED-00-23 (October 27, 1999).

losses sustained by deposit insurance programs. Cleaning up the thrift crisis in the 1980s and early 1990s cost about $150 billion.[31] Despite the implementation of risk-based premiums for banks and thrift institutions, few of them pay any deposit insurance premiums today because the government has already attained its mandated reserve levels in the deposit insurance trust funds.[32] Those examples show that the inevitable uncertainty about actuarial risks leaves room for discretion in setting prices, which in practice often leads to bigger subsidies. Accordingly, a federal disaster reinsurance program could turn out to be more costly and to encourage more risk taking than proponents claim. Moreover, as federal crop insurance has demonstrated, the government is unable to withhold disaster assistance from those at risk who fail to purchase insurance.[33]

Federal reinsurance could have three other disadvantages. First, it might crowd out private insurance suppliers. Capital market products, especially catastrophe bonds whose principal and interest would be suspended if a disaster occurred, would be at a competitive disadvantage against a subsidized federal program. That is because the cost of the capital market products would be higher than the premiums that the government would charge for its reinsurance.

Second, a federal program that nationalized disaster risk might reduce diversification gains from international risk sharing by retaining all residual risk domestically. The current system has an advantage in that reinsurance markets are international. Extensive diversification of risk contributed to the ability of insurers and reinsurers to survive the September 11 attacks.

Third, a federal reinsurance program for disasters could transfer risk and wealth in unintended directions. The rest of the nation might end up subsidizing property owners in California, New York, and Florida, the states that are most prone to catastrophic losses. Relatively affluent property owners in coastal areas and the owners of commercial properties that were landmarks would also benefit from taxpayer support.

## Other Federal Policies to Promote the Supply of Insurance

Some analysts do not think that the government needs to provide reinsurance for natural disasters or acts of terrorism. Instead, they would rely on the market to offer an adequate supply of insurance. Short of providing reinsurance, the federal government could adopt measures to encourage private supply following catastrophic events. Those measures include offering incentives for the mitigation of risks, reducing federal assistance after disasters, changing the tax treatment of loss reserves held by insurers, and limiting damage awards. To achieve some of those same goals, states could deregulate their insurance markets.

### Rely on the Market to Provide Disaster Insurance

The September 11 attack on the World Trade Center alarmed people who set financial policy because it threatened the economy rather than just the insurance industry.[34] The lack of terrorism insurance could retard commercial construction and be a drag on the economy, but

31. See Congressional Budget Office, *The Economic and Budget Outlook: Fiscal Years 1996-2000* (January 1995), pp. 46-48. Estimates for insured losses have changed over time for both banks and thrifts. See Timothy Curry and Lynn Shibut, "The Cost of the Savings and Loan Crisis: Truth and Consequences," *FDIC Banking Review*, vol. 13, no. 2 (2000), pp. 26-35.

32. Moreover, deposit insurance does not attempt to price its line of credit with the Treasury or its catastrophe coverage. See James A. Wilcox, "Issues in Deposit Insurance Reform" (paper presented at the Federal Reserve Bank of Chicago's 2001 Annual Bank Structure Conference, Chicago, May 11, 2001). Also see Joe Peek and James A. Wilcox, "Safety Net Subsidies in Banking: Decline, Resurgence, and Prospects" (working paper, University of California at Berkeley, June 24, 2001).

33. The 1996 farm bill allowed producers to opt out of catastrophe coverage provided that they signed a waiver disqualifying them from any future disaster payments for crop losses. Despite that provision, emergency relief packages in 1999 and 2000 provided assistance to farmers who had opted out of the subsidized federal crop insurance program. For further analysis, see Jerry R. Skees, "The Bad Harvest: Crop Insurance Reform Has Become a Good Idea Gone Awry," *Regulation*, vol. 24, no. 1 (Spring 2001), pp. 16-21.

34. Congressional Budget Office, *Federal Reinsurance for Terrorism Risks*, CBO Paper (October 2001).

the magnitude of that risk is still not clear.[35] Indeed, a market is developing for terrorism reinsurance to bridge that coverage gap.[36]

The largest reinsurance companies are willing to take on some terrorism risks at higher premiums, but they try to avoid issuing groups of policies that might result in large concentrated losses. For example, reinsurance provider National Indemnity, a subsidiary of Berkshire Hathaway, covers some terrorism risks for a South American refinery, several international airlines, a large North Sea oil platform, and the Sears Tower in Chicago.[37] Other reinsurers also offer terrorism coverage for airlines.[38]

In the aftermath of September 11, banks and other lenders are still providing credit on commercial projects that do not have terrorism coverage, although they may demand terrorism insurance for new structures that are landmarks in major cities.[39] A March 2002 survey of bank lending practices indicated that most banks had not tightened their loan requirements for landmark properties and other high-profile commercial real estate, including sports and entertainment venues and large shopping malls.[40] Moreover, the survey showed little decline in demand for those loans because of the lack of terrorism coverage. Most banks require terrorism insurance for less than 10 percent of their high-profile properties.

Federal and state bank regulators have not yet issued any guidelines on terrorism coverage, and bankers are tolerant of the fact that the lack of coverage may have caused some borrowers to breach their loan agreements.[41] Calling in loans is not a viable option for the banks, but they can raise fees or require more collateral.

The available data on commercial construction also do not indicate a dire situation at present. As a result of the most recent recession, which began in March 2001, construction was already slowing prior to September 11, and the current slowdown in activity has been less severe than it was during the 1990-1991 recession. Moreover, national data do not show any pronounced effects from the lack of terrorism insurance.[42] Any impact could be slow to appear, however, and spending to date may simply reflect construction that was started before September 11 and that was bolstered by mild weather in January.

35. Alan Greenspan, chairman of the Federal Reserve Board of Governors, testified in February that to date he had not seen in the data a significant impact on the economy from the lack of terrorism insurance. But he also believed that it was too early to judge. Anecdotal evidence suggests that there have been some effects. See the transcript of Greenspan's exchange with Representative Paul Kanjorski during hearings before the House Committee on Financial Services (February 27, 2002).

36. Christopher Oster, "Insurers Add Terrorism Coverage," *Wall Street Journal*, April 15, 2002, p. C18.

37. In November 2001, Berkshire Hathaway had announced that it was not going to issue terrorism insurance. But high premiums may have induced the firm to assume those risks, even if they could not be accurately priced. See the annual letter of Warren E. Buffett, Chairman of the Board, Berkshire Hathaway, to Shareholders, February 28, 2002, especially pp. 6-12, available at www.berkshirehathaway.com/letters/2001pdf.pdf.

38. Joseph B. Treaster, "A Call for the U.S. to Get Out of Terror Insurance for Airlines," *New York Times*, February 26, 2002, p. C8.

39. Real estate lending by banks has continued to increase over the past year in contrast to most other forms of bank lending, which have shown greater effects from the economic slowdown. See Board of Governors of the Federal Reserve System, "Assets and Liabilities of Commercial Banks in the United States," *Federal Reserve Statistical Release H.8* (May 3, 2002).

40. The survey asked a series of questions about how the lack of terrorism insurance had affected lending. For most of the 56 domestic banks that responded, less than 10 percent of their commercial real estate loans involved buildings or sites that are landmarks. See Federal Reserve Board, "The April 2002 Senior Loan Officer Opinion Survey on Bank Lending Practices" (April 2002), available at www.federalreserve.gov/boarddocs/snLoanSurvey/200205/default.htm.

41. See Joseph B. Treaster, "Rethinking Dire Warnings by Insurers After September 11," *New York Times*, February 27, 2002, p. C1; and Michelle Heller, "No Terrorism Insurance, But Lenders Still Lending," *American Banker*, January 7, 2002.

42. Bureau of the Census, "March 2002 Construction at $874.0 Billion Annual Rate" (press release, Washington, D.C., May 1, 2002), available at www.census.gov/ftp/pub/const/C30/c300203.pdf. The excess supply of office space in many cities has also played a role in the slowdown of commercial construction. See Federal Deposit Insurance Corporation, "Weak Fundamentals for U.S. Office Markets" (press release, Washington, D.C., March 21, 2002).

Data on commercial mortgage-backed securities (CMBSs) do not suggest a major disruption in the supply of commercial mortgage money at this time. Issuances through July 18, 2002, were down by about 20 percent from the same period in 2001, but the economic slowdown would account for most of that decline. Through July 17, 2002, spreads of CMBSs over five-year AAA swaps had actually fallen since the last week of 2001, when terrorism insurance was still widely in place. Spreads are also lower than those prevailing before September 11, 2001.[43] That small change in the spreads suggests that investors are not demanding much of a premium for bearing terrorism risk, in part because CMBSs are geographically diversified.

In spite of the market's resilience after September 11, some analysts doubt that it will be able to offer sufficient quantities of terrorism insurance in the absence of a federal backstop. The General Accounting Office found several examples of projects that have been canceled or delayed, partly because of the lack of terrorism coverage.[44] In addition, at least one bond-rating agency is considering what impact the lack of terrorism insurance may have on ratings for CMBS deals. Large commercial mortgage loans may need layers of reinsurance from several providers because no single reinsurer wants all of the risk exposure.[45] With the supply of terrorism insurance low and prices for coverage high, the Administration argues that the effect on businesses of not being able to buy insurance will be just like a tax increase on productive capital, which will ultimately be borne by workers and consumers.[46]

### Provide Incentives for Risk Mitigation

Preventive measures taken before disaster strikes can reduce the risk of loss from catastrophes. For example, developers can concentrate construction in areas less prone to catastrophic losses. To diminish losses from earthquakes, builders can bolt the wood frame of a house to its foundation and add bracing to walls.[47] In addition, securing gas water heaters to walls and installing automatic gas shutoff valves decrease the likelihood of fire after an earthquake. Homeowners in coastal regions subject to hurricanes can reinforce their roofs and install storm shutters to lessen the damage from a disaster.

But despite the potential savings that mitigation offers, most homeowners, businesses, and state and local governments do not adopt cost-effective measures to reduce their losses before a disaster. Surveys indicate that only 10 percent of the residents in areas hit by two recent California earthquakes had invested in loss-reduction measures.[48] To help prevent future losses, California has put

43. The year-to-date volume of CMBS issues was $29.2 billion as of July 18, 2002, compared with $36.6 billion for the same period in 2001. Spreads over the five-year AAA swaps increased from 42 basis points before September 11 to a high of 58 basis points in November 2001. However, the spread was just 38 basis points as of July 17, 2002, a drop of 10 basis points since the beginning of the year. Data on commercial mortgage-backed securities are available at www.cmalert.com/Public/MarketPlace/MarketStatistics/index.cfm.

44. See the statement of Richard J. Hillman, Director, Financial Markets and Community Investment, General Accounting Office, before the Subcommittee on Oversight and Investigations of the House Committee on Financial Services, published as General Accounting Office, *Terrorism Insurance: Rising Uninsured Exposure to Attacks Heightens Potential Economic Vulnerabilities*, GAO-02-472T (February 27, 2002); and Joint Economic Committee, *Economic Perspectives on Terrorism Insurance* (May 2002), available at www.house.gov/jec/.

45. Daniel B. Rubock and Tad Philipp, *CMBS: Moody's Approach to Terrorism Insurance for U.S. Commercial Real Estate*, Structured Finance, Special Report (New York: Moody's Investors Service, March 1, 2002).

46. Statement of Mark J. Warshawsky, Deputy Assistant Secretary for Economic Policy, Department of the Treasury, before the Subcommittee on Oversight and Investigations of the House Committee on Financial Services, February 27, 2002.

47. See Patricia A. Grossi, Paul Kleindorfer, and Howard Kunreuther, *The Impact of Uncertainty in Managing Seismic Risk: The Case of Earthquake Frequency and Structural Vulnerability*, Working Paper No. 99-23 (Philadelphia: Wharton Financial Institutions Center, December 1998); and Patricia A. Grossi, *Assessing the Benefits and Costs of Earthquake Mitigation*, Working Paper No. 99-24 (Philadelphia: Wharton Financial Institutions Center, June 1999). Both are available at http://fic.wharton.upenn.edu/fic/.

48. Howard Kunreuther, "Managing Catastrophic Risks Through Insurance and Mitigation" (paper presented at the Fifth Alexander Howden Conference on Disaster Insurance, Gold Coast, Australia, August 1997).

in place a program that provides subsidized loans to homeowners to retrofit their homes.

Stronger state and local building codes and better enforcement of those codes may also provide net benefits. But stronger codes are often resisted by builders, who argue that loss-prevention measures will not be appreciated by buyers and thus not fully reflected in a home's value.[49] One study estimated that 25 percent of insured losses from Hurricane Andrew could have been prevented through better building code compliance and enforcement.[50] Both Florida and California have recently adopted tougher building codes. For example, Orange County in California now requires that a new roof have plywood over the rafters to make it stronger. Also, the city of Los Angeles no longer allows masonry chimneys; manufactured chimneys, which have a wooden or stucco frame with a metal cylinder inside to vent the smoke, are acceptable.

The federal government could reform its own insurance programs to encourage risk mitigation and discourage construction in high-risk areas. By subsidizing flood insurance, for instance, the federal government has encouraged building in hazardous areas, putting more real estate at risk and raising the cost of postdisaster assistance.[51] Federal insurance will cover repeated losses (up to $350,000 for property and contents) on the same oceanfront property without an increase in premiums. The federal government could also provide financial incentives to state and local governments and private property owners to undertake measures to reduce losses.

**Reduce Postdisaster Assistance**

The federal government provides extensive financial assistance to individuals and state and local governments in areas hit by a disaster. For example, homeowners are eligible for low-interest loans from the Small Business Administration following disasters, and they can deduct uninsured property losses from taxable income.

The large share of federal disaster assistance targeted toward state and local governments is usually provided through a supplemental appropriation, which generally covers 75 percent to 100 percent of public rebuilding costs.[52] In an effort to lessen state and local governments' reliance on federal aid, which reduces their incentive to adopt mitigation measures and to purchase insurance, the Administration has proposed reducing the federal share of costs to 50 percent and requiring public buildings to carry disaster insurance.[53]

**Change the Tax Treatment of Capital Reserves**

Altering the way in which expected catastrophic losses and income from investments are taxed could expand the availability and lower the cost of disaster insurance.[54] The tax code currently limits the amount of capital reserves that insurers can accumulate by taxing as income those profits that are temporarily inflated when disaster losses are low. Insurers cannot count as expenses for tax purposes additions made to their reserves against low-

49. Michael Quint, "A Storm Over Housing Codes: Builders Resist Insurers' Call for Tougher Standards," *New York Times*, December 1, 1995, p. D1.

50. The estimate from the Insurance Information Institute is cited in Howard Kunreuther, *Mitigating Disaster Losses Through Insurance*, Working Paper No. 95-09-02 (Philadelphia: Wharton Risk Management and Decision Processes Center, 1995).

51. For an analysis of the federal government's flood insurance program and changes made to reduce its losses, see the statement of Stanley J. Czerwinski, Director, Physical Infrastructure Issues, General Accounting Office, before the Subcommittee on Housing and Community Opportunity of the House Committee on Financial Services, published as General Accounting Office, *Flood Insurance: Information on the Financial Condition of the National Flood Insurance Program*, GAO-01-992T (July 19, 2001).

52. See Rutherford H. Platt, "Shouldering the Burden: Federal Assumption of Disaster Costs," in Platt, ed., *Disasters and Democracy: The Politics of Extreme Natural Events* (Washington, D.C.: Island Press, 1999); and David A. Moss, "Courting Disaster? The Transformation of Federal Disaster Policy Since 1803," in Kenneth A. Froot, ed., *The Financing of Catastrophe Risk* (Chicago: University of Chicago Press, 1999), pp. 307-355.

53. William Claiborne, "Disaster Management Cuts Raise Concerns," *Washington Post*, May 8, 2001, p. A21.

54. Scott E. Harrington and Greg Niehaus, "Government Insurance, Tax Policy, and the Affordability and Availability of Catastrophe Insurance," *Journal of Insurance Regulation*, vol. 19, no. 4 (Summer 2001), pp. 591-612.

probability risks.[55] Instead, insurers can expense the cost of reserves only for losses that have been reported by policyholders but that remain unsettled and losses that are anticipated to have already occurred but for which claims have not yet been reported.[56]

For example, an insurer who will experience a $100 million loss from a disaster once over the next 100 years is incurring an expected loss of $1 million annually. The firm pays taxes each year on the premiums it collects until the disaster occurs, after which the firm is able to deduct the losses for tax purposes. Some of that 100-year loss may be offset by current income from other sources, and the remainder may be carried forward or back. But the potential for the firm to lose money still exists. Even if the total loss is offset or carried forward, the firm will lose the time value of its money because the taxes were paid before the deduction could be taken. Consequently, one might argue that insurers should be able to treat expected losses as a cost of doing business each year, regardless of whether insurers set aside reserves.

As an alternative, the government might allow insurance companies to put money aside tax-free to cover expected losses. State-sponsored plans generally can already do so and avoid paying taxes on investment income. Similarly, private mortgage guaranty insurers are permitted to hold 50 percent of their premiums in a tax-deductible reserve for 10 years. (The reserve must be invested in special non-interest-bearing Treasury bonds.) Another possibility is to allow insurers to carry back catastrophe losses 10 years or 20 years, instead of the current two years, and recover federal taxes paid in past years. A precedent for that alternative exists because product liability insurers can carry back losses 10 years.

Such changes to the tax code could have several disadvantages, however. First, they could lead to substantial distortions in the allocation of capital, especially between insurers and other financial institutions not subject to the same tax treatment. Second, changes would result in losses of federal revenue.[57] Third, without adequate oversight, insurers could abuse the option to set aside tax-free reserves by overestimating expected losses. Insurers and other corporations can already carry back losses two years and carry forward losses 20 years, which helps smooth out their cash flow following a disaster. Finally, altering the tax code would not address the problem that underlies insurers' reluctance to provide coverage for terrorism risk: the inability to assess risk well enough to price policies.

The taxation of income on capital reserves increases the premiums insurers must charge for catastrophe insurance.[58] Taxation of portfolio income is not unique to catastrophe insurance; it is characteristic of the corporate income tax. But catastrophe insurance is distinguished from other insurance by its necessarily high ratio of capital to expected losses and therefore by the high ratio of taxable investment income to expected losses. Consequently, insurers must increase premiums to cover those taxes and provide equity investors with at least the same expected after-tax return as they could achieve by purchasing a similar investment directly. For instance, say the insurer in the previous example earns 6 percent on its $100 million bond portfolio, which backs the risk of catastrophic loss. If taxes absorbed one-quarter of the investment income, then the insurer's 6 percent market return on its investments would be reduced to 4.5 percent after taxes. Thus, the insurer would need to collect an additional 1.5 percentage points ($1.5 million) in premiums beyond the $1 million in expected losses to compensate capital providers for catastrophe exposure.

55. Proponents of this tax treatment argue that insurers would opportunistically set aside too large a portion of their reserves for catastrophes if reserves were allowed to accumulate tax-free. See Robert E. Litan, "Earthquake! Planning and Paying for the 'Big One'," *Brookings Review* (Fall 1990), pp. 42-48.

56. David F. Bradford and Kyle D. Logue, *The Influence of Income Tax Rules on Insurance Reserves,* Working Paper No. 5902 (Cambridge, Mass.: National Bureau of Economic Research, January 1997).

57. The Joint Committee on Taxation estimated that establishing tax-free reserves for terrorism coverage would reduce revenues by $12.4 billion over 10 years. See Congressional Budget Office, *Cost Estimate for H.R. 3210, Terrorism Risk Protection Act* (November 19, 2001).

58. Scott E. Harrington and Greg Niehaus, "On the Tax Costs of Equity Finance: The Case of Catastrophe Insurance" (working paper, University of South Carolina, May 2001).

That tax treatment may help explain the limited scope of catastrophe insurance and the growth of untaxed reinsurers incorporated in Bermuda. But only highly complicated alterations of the corporate tax code could change them. The same problem arises with respect to other competing financial institutions but to a different degree. An analysis of tax-code changes is beyond the scope of this paper, however.

**Limit Damage Awards**

Another way to increase the viability of privately provided insurance is to limit the liability of companies, and therefore insurers, for losses from terrorist attacks, as H.R. 3210 does. Such restrictions could stand alone or be combined with other policies. For instance, the federal government could limit the maximum awards for third-party damages or the awards for economic loss and pain and suffering associated with death benefits.

When considering limits on damage awards, policymakers need to balance the goals of making terrorism insurance available and affordable and of providing firms with incentives to mitigate risks.[59] Restricting damage awards would make risk more manageable for insurers and firms.[60] Although considerable uncertainty would remain about the probability of attacks and the number of people likely to be affected, the restrictions would lessen some of the uncertainty associated with the potential size of claims and could reduce the likelihood of litigation. Any such limitations would increase the risk borne by potential claimants and could reduce firms' incentives to mitigate losses.[61]

**Deregulate Insurance Markets**

As another way to promote the supply of insurance, the federal government could encourage states to deregulate their insurance markets, giving consumers more choices. Many analysts argue that no strong economic rationale exists for regulating insurance companies. In particular, the structure, conduct, and performance of the homeowners' insurance industry is already fairly competitive. Moreover, there is no clear evidence of firms earning excess profits or exercising other forms of market power.[62]

Deregulation offers benefits as well as drawbacks. Deregulating rates and allowing insurance firms to leave Florida as well as ending the mandatory offer of earthquake insurance in California could heighten companies' earnings prospects, increase their ability to purchase reinsurance, and encourage entry into the market. Deregulation would also reduce the likelihood of a federal reinsurance program having to pay for mistakes made by state regulators. On the downside, consumers would pay more for disaster insurance in Florida, and, to some extent, in California, at least in the short run. In addition, insurance markets would still be subject to disruption, and companies would be prone to insolvency, after a catastrophe.[63]

---

59. For an analysis of the trade-offs in the context of environmental liabilities, including a discussion of the Price-Anderson Act, which limits the nuclear industry's liability, see James Boyd, *Financial Responsibility for Environmental Obligations: Are Bonding and Assurance Rules Fulfilling Their Promise?* Discussion Paper 01-42 (Washington, D.C.: Resources for the Future, August 2001).

60. See Allan M. Feldman and John Frost, "A Simple Model of Efficient Tort Liability Rules," *International Review of Law and Economics*, vol. 18, no. 2 (June 1998), pp. 201-215; and James Boyd and Daniel Ingberman, "Do Punitive Damages Promote Deterrence?" *International Review of Law and Economics*, vol. 19, no. 1 (March 1999), pp. 47-68.

61. For a review of the importance of liability rules on loss prevention and the maximization of social welfare, see I.P.L. Png, "Litigation, Liability, and Incentives for Care," *Journal of Public Economics*, vol. 34, no. 1 (October 1987), pp. 61-85.

62. Scott E. Harrington, *Insurance Deregulation and the Public Interest* (Washington, D.C.: AEI-Brookings Joint Center for Regulatory Studies, 2000).

63. J. David Cummins, Neil Doherty, and Anita Lo, *Can Insurers Pay for the 'Big One'? Measuring the Capacity of the Insurance Industry to Respond to Catastrophic Losses*, Working Paper No. 98-11-B (Philadelphia: Wharton Financial Institutions Center, June 24, 1999), available at http://fic.wharton.upenn.edu/fic/papers/98/9811.pdf.

APPENDIX

# A

# State-Sponsored Insurance Programs in Florida and California

State-sponsored programs provide natural disaster insurance at low prices directly to policyholders at the greatest risk. They also offer reinsurance to companies that issue coverage for natural disasters. (No coverage is provided for terrorism risks.) In some cases, the programs only cover risks rejected by private insurers. Rates are generally subsidized, although not explicitly, and subsidies vary across different types of policyholders.

The ability of state-sponsored programs to offer low-cost insurance is more illusory than real, for several reasons. First, most of the programs appear to be undercapitalized. Major catastrophic events would probably result in deficits for the programs, which would then need to be funded through assessments on all policyholders. A catastrophe would also significantly increase a state's debt. Although states are not legally required to back the insurance pools, political considerations make such support very likely following a disaster. Second, today's low premiums are in effect being paid for by premiums and taxpayer liabilities that will in all likelihood be higher in the future. Low premiums have the added disadvantage of discouraging preventive measures that would reduce losses over the long term. Third, because state-sponsored plans generally base postdisaster assessments on an insurer's market share, they discourage private-sector insurers from issuing new policies.

## Florida Residential Property and Casualty Joint Underwriting Association

All property insurance companies underwriting policies in Florida are required to belong to the Florida Residential Property and Casualty Joint Underwriting Association (FRPCJUA), which primarily provides residential coverage to property owners who were unable to obtain coverage from private insurers after Hurricane Andrew struck in 1992. (In certain high-risk areas, separate windstorm coverage is provided by the Florida Windstorm Underwriting Association.) Rates charged by the FRPCJUA in each county must be at least as high as the highest rates charged by the 20 largest private insurance companies in Florida. Despite that requirement, however, the association's rates are still likely to be below market rates because the association faces adverse selection: any property owner who could find cheaper coverage in the private market would do so.

If the association runs a deficit, as it did in 1994 and 1995, it can levy assessments—up to 10 percent of the deficit or 10 percent of statewide premiums, whichever is greater—on each member on the basis of its market share in the state. The insurance companies can then pass through the cost of the assessments to their policyholders by filing for a recoupment. If the special assessments are insufficient to eliminate the deficit, the association can issue bonds, which will be paid off through emergency assessments directly levied on all policyholders of property and casualty insurance. The association has operated at a profit since 1996, but no major loss has occurred since Hurricane Andrew.

The association's capacity to pay claims was $1,864 million as of January 31, 2002. That amount greatly exceeds the probable maximum loss that the association would face from a disaster that occurred once every 100 years, on average, but it is less than the loss from a disaster that occurred once every 250 years.[1] The association would pay claims first by drawing down its surplus of $190 million. Once that surplus was exhausted, the association would pay claims through private reinsurance ($34 million), assessments on members ($400 million), reimbursement from the state-sponsored Florida Hurricane Catastrophe Fund ($320 million), pre-event disaster notes ($350 million), and a line of credit ($570 million). Thus, most of the association's payment capacity is based on special assessments, borrowed funds, and reinsurance purchased from another state program.

For several years, the FRPCJUA was the second-largest insurer in Florida. The association's exposure peaked in the fall of 1996 at $98 billion, when it had 937,000 policies. To encourage private-market coverage, the Florida legislature granted various incentives—cash payments and exemptions from special assessments—to firms taking policies out of the fund. Those efforts have been successful; as of June 2002, the association had $19.3 billion in exposure from roughly 119,000 policies.[2] Private companies are allowed to remove policies from the pool by selectively matching its rates and terms, thus leaving the association with only the most underpriced policies.

But the success of the effort to transfer risk from the association to private insurers may prove temporary. Rating agency A.M. Best has expressed concern that the incentives offered by the state legislature encouraged financially weak insurers to take policies out of the fund. Those firms may become financially impaired or insolvent from a low-intensity hurricane. Moreover, their thin capitalization has led them to purchase 90 percent catastrophe reinsurance from the Florida Hurricane Catastrophe Fund. In effect, those insurers may be making a one-sided bet. If no hurricanes hit Florida, they gain; if one hits, both the Florida Guaranty Fund, a state-sponsored fund that pays off claims for the state's insolvent insurers, and the hurricane fund lose.[3]

## Florida Windstorm Underwriting Association

The Florida Windstorm Underwriting Association (FWUA) provides hurricane and windstorm coverage to property owners who cannot obtain that coverage from private insurance companies. Standard homeowners' policies from private insurers must cover hurricane losses in most areas, but insurers may exclude those losses in areas covered by the association. By taking on hurricane risk, the FWUA makes the prospect of offering standard homeowners' policies in high-risk areas more attractive to private insurers, thus reducing pressure on the FRPCJUA to provide coverage.

Created by the state legislature in 1970 to offer coverage in the Florida Keys, the FWUA provides coverage to the most exposed areas in 29 of Florida's 35 coastal counties. (Coverage is more extensive in south Florida than in the rest of the state.) In some cases, only properties within a quarter-mile of the beach are eligible for coverage.

The median value of a home in areas of the state that are eligible for coverage is almost 70 percent higher than in areas that are not eligible, suggesting that poorer property

1. The modeling firm EQECATestimated that a 100-year loss would probably cost the association a maximum of $1.2 billion, and a 250-year loss would cost the association $2.1 billion. See Florida Residential Property and Casualty Joint Underwriting Association, "JUA Resources to Pay Hurricane Claims as of January 31, 2002," available at www.frpcjua.com/WWWjua.nsf/VURL/Claims.

2. Ron Bartlett, "Bringing Stability to a Market in Crisis: An Overview of the FRPCJUA, 1992-2000" (paper, Florida Residential Property and Casualty Joint Underwriting Association, May 2000), available at www.frpcjua.com (JUA Information). Figures have been updated using the latest "Policies & Exposure Report." The number of policies reached a low of just under 60,000 in April 2000 with a total exposure of $9.5 billion.

3. A.M. Best Company, "Florida Insurers May Be Unprepared for Major Storms," *Best's Viewpoint,* Release 6 (March 13, 2000).

owners could end up cross-subsidizing richer ones through assessments imposed after a disaster.[4] Special assessments on members were imposed after two hurricanes in 1995—an assessment of $84 million after Hurricane Opal and one of $33 million after Hurricane Erin. (Each of those hurricanes caused insured losses of less than $1 billion in Florida.) A $100 million assessment followed Hurricanes Georges and Mitch in 1998. Those assessments were borne by all property and casualty policyholders in the state.

The FWUA currently has almost 410,000 policies with an exposure exceeding $98 billion; annual premiums and surcharges approach $414 million.[5] Before Hurricane Andrew, the FWUA had fewer than 62,000 policies and less than $8 billion of exposure. Much of the association's growth stemmed from decisions in 1997 and 1998 by State Farm and Allstate, the two largest private insurers in Florida, to stop offering windstorm coverage to a large number of policyholders.

Although the basic operating principles of the Florida Windstorm Underwriting Association are similar to those of the FRPCJUA, the two associations differ in the way they set their rates. Under state law, the FRPCJUA's rates must equal or exceed the highest rates charged by the state's 20 largest insurance carriers. The Windstorm Underwriting Association, in contrast, must set its rates through the regulated process that applies to rate filings in general. Because the FWUA covers wind perils for which computer modeling may be the best available tool to determine rates and because the association's board is composed of representatives from private insurers, rate setting for the FWUA has been contentious. For example, on the basis of its computer modeling, the association asked for a 61 percent rate hike in 1997 (*see Table A-1*). That request was turned down by the Florida Department of Insurance; on appeal, arbitrators awarded a 12 percent increase. In May 1999, the association requested a 96 percent increase to be phased in over several years. That request was also denied by regulators, but arbitrators overturned the decision during the appeals process. The arbitration decision is now being challenged in the courts.

Rate comparisons based on data filed with the Florida Department of Insurance consistently show that the Windstorm Underwriting Association's rates are significantly below rates for private policies in the same areas. In Dade County, for example, State Farm charges an average of $1,260 a year for hurricane coverage for a $150,000 home. The same coverage in the FWUA plan costs just over $880.[6] Rates are also below expected annual losses, even though Florida law mandates that rates be actuarially sound.[7]

As of February 2002, the FWUA had a claims-paying capacity of $5.8 billion.[8] A storm that occurred an average of once every 100 years could cost the association $4.8 billion and necessitate more than $1 billion of regular and emergency assessments.[9] (If the FWUA incurs a deficit, it can impose regular assessments, the amount of

4. Rade T. Musulin, "Would a Federal Role in Disaster Protection Be a Catastrophe?" *Contingencies*, American Academy of Actuaries (November/December 2000).

5. See FWUA's "Inforce Liabilities and Policies Summary Spreadsheet as of January 31, 2002," available at www.fwua.com (status reports).

6. The differences in coverage for screened-in enclosures and other detached dwellings complicate rate comparisons between the FWUA and private insurers. Nonetheless, data show that the association's rates are significantly below market rates. See Florida Senate Committee on Banking and Insurance, "Availability and Cost of Residential Hurricane Coverage," Report No. 2000-03 (August 1999), pp. 28-31 and Appendix A; and Florida Senate Committee on Banking and Insurance, "Availability and Cost of Residential Hurricane Coverage," revised report (September 1999), p. 7.

7. Florida Senate Committee on Banking and Insurance, "Availability and Cost of Residential Hurricane Coverage," p. 4. The Florida Senate considered in 2001 a joint resolution (S300 Windstorm Insurance Rate Increases) proposing a constitutional amendment that would limit rate increases to 3 percent annually. Another bill would have prohibited the wind pool from using arbitration to raise rates and would have shifted about 200,000 homes back to the private market through an assigned-risk program.

8. FWUA, "Florida Windstorm Underwriting Association 2002 Claims Paying Resources," available at www.fwua.com/ftp/claims payingresource0202.pdf.

9. Florida Senate Committee on Banking and Insurance, "Availability and Cost of Residential Hurricane Coverage," Table 5, p. 18. A 250-year loss would cost $7.4 billion, and a 500-year loss would cost $9.5 billion as of November 30, 2000. Data were supplied to the Congressional Budget Office by the Florida Windstorm Underwriting Association in February 2002.

**Table A-1.**

## Arbitration Decisions in Florida

| Company | Filing Date | Requested Rate Change (Percent) | Arbitration Decision (Percent) | Date of Panel's Decision | Effective Date of Rate Change |
|---|---|---|---|---|---|
| State Farm Fire & Casualty | May 1997 | 25.6 | 25.6 | June 1997 | October 1997 |
| Continental Insurance Group | August 1997 | 28.1 | 0 | May 1998 | August 1998 |
| Florida Windstorm Underwriting Association | August 1997 | 61.0[a] | 12.0 | March 1998 | August 1998 |
| USAA | September 1997 | 19.4 | 14.8 | June 1998 | October 1998 |
| Nationwide Insurance Co. of Florida | December 1998 | 29.0 | 18.0 | September 1999 | July 2000 |
| Florida Windstorm Underwriting Association | May 1999 | 96.0[b] | 96.0[b] | February 2000 | Unknown[c] |
| First Floridian | June 1999 | 17.2 | 11.8 | February 2000 | Unknown[c] |
| State Farm Florida | October 1999 | 7.0 | 7.0 | June 2000 | Unknown |
| USAA | November 1999 | 16.6 | 7.7 | May 2000 | October 2000 |
| Cypress (Homeowners' Program) | February 2000 | 12.0 | 0 | September 2000 | October 2000 |
| Cypress (Dwellings) | February 2000 | 14.0 | 0 | September 2000 | October 2000 |

Source: Congressional Budget Office based on Florida Senate Committee on Banking and Insurance, *Availability and Cost of Residential Hurricane Coverage*, Report No. 2000-03 (August 1999), Table 1, p. 8. Updates were provided by Rade T. Musulin, vice president of the Florida Farm Bureau Insurance Companies.

a. This rate change was to be implemented over three years.

b. Rate filing was to be implemented with caps of 20 percent and 30 percent for years one and two, respectively. A cap of 40 percent would apply in subsequent years.

c. Arbitration decision is being challenged in court.

which is the greater of 10 percent of the association's deficit or 10 percent of property insurance premiums in the state.) As of December 31, 2001, the association had a surplus of $290 million. Like all residential insurers, the association must purchase reinsurance from the Florida Hurricane Catastrophe Fund, which would cover just over 50 percent of the cost of a 100-year storm.[10] Pre-event notes issued by the association as part of its payment capacity exceed $1.75 billion.

10. Presentation by the Florida Hurricane Catastrophe Fund before the Florida State Board of Administration's cabinet meeting, June 26, 2000.

## Florida Hurricane Catastrophe Fund

As the world's largest hurricane reinsurer, the Florida Hurricane Catastrophe Fund is used extensively by the state's other two residual pools as well as by private insurers. When created in November 1993, the fund provided reinsurance on commercial properties in addition to residential properties, but it has since stopped offering commercial coverage, partly because commercial reinsurance is widely available in the private market. (Administrative difficulties also played a role in the fund's decision to drop its commercial reinsurance program.)

Florida requires all insurers to participate in the Hurricane Catastrophe Fund. Each private insurer may choose coverage that pays 45 percent, 75 percent, or 90 percent of losses in excess of its "retention level," a deductible determined by dividing the programwide retention among individual insurers in proportion to their share of the fund's total premiums. (Both the FRPCJUA and the FWUA must purchase 90 percent coverage.) In 2001, the total retention level was about $3.4 billion. The level increases annually with the growth of premiums. To retain its tax-exempt status, the fund must invest about $10 million a year in public projects that reduce hurricane-related losses.[11]

Annual rates for residential coverage exceed a total of $475 million—or more than 13 percent of the estimated $3.5 billion in residential premiums collected in Florida in 2001. Those rates are based on hurricane models approved by the Florida Commission on Hurricane Loss Projection Methodology. The fund's rate-on-line—its ratio of premiums paid to coverage purchased—is now significantly below rates in the private market, and that gap has widened over time with growth in the fund's claims-paying capacity. During the 1995-1996 contract season, for example, the fund's rate-on-line was about 9 percent (a $1 premium bought $11 of coverage), more than twice as high as the rate-on-line of roughly 4 percent (a $1 premium bought $25 of coverage) seen in the first half of 2001. In comparison, the rate-on-line for private reinsurance coverage in Florida was closer to 10 percent (a $1 premium bought $10 of coverage) in the first half of 2001.[12] The fund's rate-on-line rises significantly after a disaster because the losses reduce its claims-paying capacity.

The fund's total exposure was $919 billion as of December 31, 2001, but the state legislature capped the claims-paying capacity of the fund at $11 billion—a cash surplus of $4.3 billion and bonding authority of $6.7 billion. Combined with the retained risks of the private insurers and state pools, the fund could have hypothetically covered a $16 billion disaster in 2000.[13] Any insured loss that exceeded $16 billion (a 100-year loss was estimated to cost $21.6 billion in 2001, and a 250-year loss was estimated at $31.9 billion) would require that claims be paid on a prorated basis. However, by paying less than the full amount on reinsurance contracts, the fund could damage the ability of all insurers, including the FRPCJUA and the FWUA, to meet their policyholders' claims. In that event, the state of Florida might have little choice but to provide additional money to the fund, even though it is not legally obligated to do so.

Whether the fund could actually borrow as much as the state has authorized is uncertain. A $16 billion disaster would require a municipal bond issue of $6.7 billion, an unprecedented amount for a state to borrow. The bond would be backed by an emergency assessment of 3.47 percent, which would be directly added to all premiums for property and casualty policies—except for workers' compensation policies—in the state. Moreover, the FRPCJUA and the FWUA might be forced to impose special levies at the same time, which could provoke a consumer backlash. The legislature also intends that the catastrophe fund rebuild its capacity through a second bond issue of nearly $7.9 billion.[14] That bond issue would be backed by an emergency assessment of 2.53 percent directly levied on all policyholders.

## California Earthquake Authority

California law requires insurers to offer earthquake coverage as a rider to homeowners' policies. Insurers can satisfy that requirement by offering coverage from the California

11. Neither the FRPCJUA nor the Windstorm Underwriting Association has tax-exempt status; however, the Florida legislature is considering some structural changes that could make the two funds tax-exempt. Personal communication to the Congressional Budget Office from Jack Nicholson, director of the Florida Hurricane Catastrophe Fund, February 25, 2002.

12. See Rade T. Musulin, "Federal Natural Disaster Legislation: Current Issues with H.R. 219" (discussion draft, Florida Farm Bureau Insurance Companies, June 13, 1998); and Musulin, "Taming the Big Cats," *Best's Review*, A.M. Best Company (March 1999). Both papers are available at www.ffbic.com/actuary/papers/.

13. The financial data and actuarial estimates come from the presentation by the Florida Hurricane Catastrophe Fund before the State Board of Administration's cabinet meeting and the fund's *Annual Report* for fiscal year 1998-1999. But actuary Rade Musulin argues that a $16 billion loss is an event that would occur once every 65 years, on average.

14. Guy Carpenter & Company, Inc., "Florida Hurricane Catastrophe Fund 1999/2000—Creating a Stable Fund," Special Report (October 1999), available at www.guycarp.com/pdf/flcatfund.pdf. Figures were updated to 2001 using data from the Florida Hurricane Catastrophe Fund.

Earthquake Authority (CEA), which was established by the state legislature following the Northridge earthquake in 1994. The CEA's coverage is less extensive than the coverage that was available privately before the Northridge quake. For example, it carries a deductible of 15 percent of the amount of insurance on a property, excludes coverage for swimming pools and detached garages, and limits coverage to $5,000 for contents and to $1,500 for additional living expenses. The CEA's premiums are risk-based; they vary depending on a home's location relative to fault zones, its type of construction, its age, its degree of retrofitting, and the type of soil underneath it. In November 2001, the average policy for a house cost about $560 a year (or $2.37 per $1,000 of coverage), but costs were several times higher in areas with greater risk of seismic activity.

Neither the government nor mortgage lenders requires consumers to purchase earthquake insurance. Only about 16 percent of homeowners in California have earthquake coverage today, down from 30 percent just after the Northridge quake.[15] Consumer interest in the CEA's policies has been limited because of coverage restrictions and other factors. To encourage more homeowners to purchase coverage, the CEA now offers policies with lower deductibles and broader coverage at higher premiums.[16]

In contrast to insurers in most other states that sponsor programs, insurers in California are not obligated to join the CEA's program. However, the CEA has approximately 66 percent of the residential market.[17] The remaining firms—now about 90—offer their own earthquake policies. But most of those firms have been able to hold down the number of residential earthquake policies they issue by charging relatively high rates and by carefully selecting the location of the homes they insure. The main deterrent to participation in the state-sponsored program is the threat of assessments on insurers and policyholders after a disaster.[18]

As of January 2002, the CEA had approximately 814,000 basic earthquake policies outstanding and an exposure from all policies of $174.7 billion; the authority's total payment capacity was $7.36 billion.[19] In addition, its annual premiums will exceed $429 million in 2002. Although the CEA has accessed the private reinsurance markets and arranged contingent lines of credit, it still relies on postdisaster assessments to meet nearly half of its maximum payout. As of November 2001, the CEA's capacity to pay claims consisted of the following layers: capital or cash reserves ($1,034 million); a first assessment on participating insurers ($2,183 million); a first layer of reinsurance provided by a consortium of private reinsurance firms ($1,433 million); a line of credit provided by a consortium of banks ($716 million); a second layer of reinsurance obtained primarily from Berkshire Hathaway ($538 million); and a second assessment on participating insurers ($1,456 million).[20] The CEA would impose surcharges on policyholders—of up to 20 percent of premiums—and on insurers to repay its debt issues. Insurers would in turn try to recoup those assessments by passing them through to their customers.

The CEA asserts that its current capacity would be adequate to pay all claims that it would face from an earthquake roughly two and a half times as costly as the Northridge quake. (Had the authority been in operation when the 1994 Northridge earthquake struck, it would have

15. Personal communication to the Congressional Budget Office from consulting actuary Richard J. Roth Jr., July 16, 2001.

16. As of January 24, 2002, almost 113,000 supplemental policies, at an average annual premium of $184, were in force with a total exposure of $7 billion, according to the California Earthquake Authority's supplemental weekly status report.

17. Personal communication to the Congressional Budget Office from Stan Devereux, Legislative and Public Affairs Director, California Earthquake Authority, January 25, 2002.

18. The number of policies issued by the California Earthquake Authority declined from 1995 to 2001, a period during which there was little seismic activity. In contrast, the amount of commercial earthquake insurance increased during that period. Analysts attribute much of that difference in trends to the lack of a mandatory offer requirement for commercial insurers. Personal communication to CBO from Richard J. Roth Jr., July 27, 2002.

19. According to CEA's basic weekly status report of January 24, 2002.

20. Personal communication to CBO from Stan Devereux, January 25, 2002.

faced claims of about $4 billion, according to estimates.)[21] Indeed, the authority contends that it could pay all its claims from a loss of that magnitude occurring once every 1,000 years, which puts the CEA's probability of default at less than 0.1 percent.[22] If claims exceeded the fund's capacity, the CEA would prorate payments. The state has no legal obligation to stand behind the CEA, but many policyholders might assume an implicit guarantee.[23]

21. That figure is lower than the $16.6 billion (in 1999 dollars) in actual claims resulting from the Northridge quake because the limited coverage of policies issued by the CEA would have resulted in fewer and smaller claims and policies would not have included commercial losses.

22. Those assertions are based on estimates of expected losses made by EQECAT, a modeling firm. A 500-year loss would result in the CEA paying about $6 billion in claims, while a 100-year loss would trigger about $3 billion in claims. Such estimates are, however, uncertain. For details on capacity and the probability of losses, see Tillinghast-Towers Perrin, "CEA Project Consulting Team Report" (New York, July 5, 2001), especially Appendix A.

23. By making the CEA independent, the state has avoided having those contingent liabilities counted against its bond ratings.

Because the residential earthquake coverage offered by the CEA is minimal, critics have suggested two reforms to open up California's market to private insurers.[24] First, repealing the law that requires insurers to offer earthquake coverage would allow them to spread policies geographically, thus reducing their risk. Second, allowing members of the CEA to also issue their own earthquake policies to some of their policyholders would put the industry's capital and surplus to better use. The top three insurers participating in the CEA have a total of $60 billion in capital and surplus, which could be used to support the market for residential earthquake insurance. An advantage of those reforms is that private insurers would have an incentive to offer earthquake policies that were more attractive than the policies currently offered by the CEA.

24. Personal communication to CBO from Richard J. Roth Jr., July 16, 2001.

# Innovations in Capital Markets

Financial markets have recently developed several new types of securities to spread catastrophic risks among capital market investors.[1] Traders use variations on standard financial instruments, such as options and bonds, to transfer income received in the form of premiums to the capital markets for their assumption of risk. The amount of risk transferred to the markets has been relatively small—roughly $2.5 billion a year between 1996 and 2000. But the development of those catastrophe-linked securities should make the reinsurance market slightly more competitive.[2] That could slow the climb in prices following a future hurricane or earthquake.[3] None of the transfers have involved terrorism risks, but those risks could, in theory, also be shared with the capital markets.

Capital markets may prove to be an efficient alternative to traditional reinsurance markets.[4] Because of their larger size, international capital markets could absorb the losses from a natural disaster without significant disruption. Indeed, daily fluctuations in the overall value of the capital markets commonly exceed the losses associated with the largest natural disasters to date.[5]

Catastrophe-linked securities also offer investors a major advantage—portfolio diversification. Losses from natural disasters are largely uncorrelated with changes in the stock and bond markets.[6] For that reason, the price of catastrophic protection should theoretically be close to insurers' expected losses, which means that such investments would deliver a return comparable to that of Treasury

1. For example, see John Pollner, *Catastrophe Risk Management: Using Alternative Risk Financing and Insurance Pooling Mechanisms*, Working Paper No. 2560 (Washington, D.C.: World Bank, February 26, 2001); Sara Borden and Asani Sarkar, "Securitizing Property Catastrophe Risk," *Current Issues in Economics and Finance*, Federal Reserve Bank of New York, vol. 2, no. 9 (August 1996); and Insurance Services Office, Inc., *Financing Catastrophe Risk: Capital Market Solutions* (Jersey City, N.J.: ISO Insurance Issues Series, January 1999).

2. For an analysis of the factors retarding growth of insurance-linked securities, see Swiss Re, "Capital Market Innovation in the Insurance Industry," *Sigma*, no. 3 (2001), available at www.swissre.com in the *Sigma* archive.

3. Kenneth A. Froot, *The Evolving Market for Catastrophic Event Risk*, Working Paper No. 7287 (Cambridge, Mass.: National Bureau of Economic Research, August 1999).

4. See Dwight M. Jaffee and Thomas Russell, "Catastrophe Insurance, Capital Markets, and Uninsurable Risks," *Journal of Risk and Insurance*, vol. 64, no. 2 (June 1997), pp. 205-230.

5. Neil A. Doherty, "Financial Innovation in the Management of Catastrophe Risk," *Journal of Applied Corporate Finance*, vol. 10, no. 3 (Fall 1997), pp. 84-95.

6. In contrast, the September 11 terrorist attacks triggered a substantial short-term drop in the stock market.

securities.[7] In practice, however, catastrophe-linked securities and derivatives appear to deliver excess returns to investors. That anomaly may result from the residual uncertainty that surrounds estimates of expected losses, the lack of liquidity in the market, and investors' high cost of information.

## Disaster Bonds

Disaster or catastrophe bonds (also known as "Act of God" bonds) forgive interest and principal, in part or in full, in the event of specified catastrophes. After a covered event, those forgiveness provisions enable the issuer—an insurer—to use the money that would have otherwise been paid to bondholders to instead pay catastrophe-related claims. Bond purchasers are compensated for those provisions by receiving a higher interest rate before a disaster strikes. (*See Box B-1* for an example.) Between 1994 and early 2002, a total of $5.5 billion in catastrophe bonds was issued in 42 transactions.[8] Most of the issues were for risks with less than a 1 percent likelihood of loss. To date, no losses have occurred on the bonds.[9]

Disaster bonds have benefits as well as costs. For the issuer, they provide better protection than reinsurance, which carries the risk of counterparty (reinsurer) default. For investors, disaster bonds provide an element of security because they are collateralized by Treasury securities. Investors are exposed only to the risk of a disaster, not to the underlying credit risk of the issuer. Currently, insurers must establish an offshore special-purpose reinsurer to issue catastrophe bonds and then issue the reinsurance. That structure allows primary insurers to treat the catastrophe bonds as reinsurance rather than debt for tax and accounting purposes, but it also significantly raises the cost of issuing the bonds.

If the estimates of losses generated by catastrophe models are correct, then people who invest in disaster bonds are earning excess returns.[10] A survey of transactions over the past several years showed that the yield on catastrophe bonds, on average, was nearly seven times the expected loss. Investors apparently require high yields to compensate for several factors: the bonds' lack of liquidity, the possibility that expected losses may be higher than estimated, and the cost of educating themselves about the bonds.

The high yield on disaster bonds may also result in part from market perceptions of moral hazard. Even though insurers cannot control the likelihood of a disaster, catastrophe bonds may elevate insurers' tolerance for risk. As a result, insurers may increase the number of policies issued in areas prone to disasters and be less aggressive in pushing mitigation and in managing claims for losses beyond the deductible.[11] Those actions could cause insurers' losses to be higher than expected, lowering returns to investors.

---

7. In theory, investors would expect to earn the risk-free rate of return when they were not exposed to the credit risk of the underlying firm and the catastrophic losses had zero correlation with market returns.

8. See J. David Cummins and Christopher M. Lewis, "Advantages and Disadvantages of Securitized Risk Instruments as Pension Fund Investments" (paper presented at the Pension Research Council and Wharton School of the University of Pennsylvania Conference on Risk Transfers and Retirement Income Security, Philadelphia, April 23, 2002). Only part of the bond issue represents a transfer of risk because only part of the bond's principal and interest are at risk. See J. David Cummins, David Lalonde, and Richard D. Phillips, *The Basis Risk of Catastrophic-Loss Index Securities* (Philadelphia: Wharton Financial Institutions Center, May 24, 2000); General Accounting Office, "Insurers' Ability to Pay Catastrophe Claims," GAO/GGD-00-57R (February 8, 2000); and Joe Niedzielski, "Catastrophe-Bond Market Is Poised for Growth in Wake of Significant Storm Damage Last Year," *Wall Street Journal*, June 12, 2000, p. C29.

9. Lehman Brothers, "Review and Application of Capital Market Products," Appendix A of Tillinghast-Towers Perrin, "CEA Project Consulting Team Report" (New York, July 5, 2001), p. A-24.

10. Vivek Bantwal and Howard Kunreuther, *A Cat Bond Premium Puzzle?* Working Paper No. 99-05-10 (Philadelphia: Wharton Financial Institutions Center, May 28, 1999).

11. Neil A. Doherty, "Financial Innovation for Financing and Hedging Catastrophe Risk" (paper presented at the Fifth Alexander Howden Conference on Disaster Insurance, Gold Coast, Australia, August 1997).

**Box B-1.**

## USAA's Catastrophe Bonds

In 1997, a special-purpose reinsurer (an insurer who assumes all or part of another insurer's risk for unusual hazards) issued a $400 million catastrophe bond for USAA, a large mutual insurance company owned by its policyholders. In effect, USAA was purchasing reinsurance coverage for 80 percent of its losses between $1 billion and $1.5 billion from a single event.[1] Although traditional reinsurance probably was available at a lower cost, the firm's goal with the bond issue was to gain access to another source of risk-bearing capacity, one that might lead to lower prices for coverage and a greater supply. The deal took months to develop. Rating agencies needed to establish criteria for rating the bonds, and regulators had to agree that the bondholders were not writing contracts for, or issuing, insurance.

The price of such a bond, whose rate of return is largely independent of other financial assets, would be expected to equal the risk-free rate plus 63 basis points (a basis point is one-hundredth of a percentage point), the expected loss from a catastrophe.[2] However, people who invested in the 1998 issue received 576 basis points over the London interbank offered rate (LIBOR), or nine times the expected loss.[3] The bond issues were significantly oversubscribed, and secondary-market yields, as measured in basis points over LIBOR, fell to the mid-400s. The substantial premiums paid to place the catastrophe bonds may be viewed in two ways: as excessive or as a development cost necessary to establish that market and educate investors.

Premiums fell on subsequent USAA bond issues, as did estimates of expected losses, but those declines resulted from incremental changes to the forecast model. People who invested in the 1999 bond issue, which was for $200 million, received 366 basis points over LIBOR to cover an expected loss of 44 basis points. As a result, the ratio of premiums to expected losses was still more than 8 to 1 in 1999.[4] USAA also purchased nearly $250 million of traditional reinsurance that year at nearly identical terms; thus, pricing in the two markets was similar.

1. For more details on USAA's various bond issues, see Kenneth A. Froot, *The Market for Catastrophe Risk: A Clinical Examination*, Working Paper No. 7286 (Cambridge, Mass.: National Bureau of Economic Research, February 2001).

2. Ibid., pp. 9-14.

3. Uncertainty over the actuarial estimates of loss also affects the pricing of catastrophe bonds. See James F. Moore, *Tail Estimation and Catastrophe Security Pricing: Can We Tell What Target We Hit If We Are Shooting in the Dark?* Working Paper No. 99-14 (Philadelphia: Wharton Financial Institutions Center, 1999).

4. Froot, *The Market for Catastrophe Risk*, p. 14.

## Catastrophe Options

Exchange-traded catastrophe options give the holder a right to payment if an index of catastrophe-related losses exceeds a specified threshold. Because the options are standardized, they can be traded at short notice and with little cost—an advantage over other forms of insurance. Catastrophe options are similar to the excess-of-loss reinsurance contracts that have been proposed for sale by the federal government.[12] The options that were traded on the Chicago Board of Trade until December 2000, when trading in those options ceased, protected insurers from total insured losses of up to $50 billion.[13] However, trad-

12. An option is the right, but not the requirement, to buy an asset at a fixed price. Most trades of catastrophe options create call spreads—a trading strategy in which a market participant simultaneously buys a call at one strike price and sells another call at a higher strike price, with both calls expiring on the same date. That strategy is basically equivalent to buying a layer of reinsurance. See Borden and Sarkar, "Securitizing Property Catastrophe Risk."

13. Between September 1995 and April 1998, about $80 million in reinsurance capacity was created through the Chicago Board of Trade. The market grew by 65 percent between 1996 and 1997. See the statement of Sylvia Bouriaux, Group Manager, Financial Products, Chicago Board of Trade, before the House Committee on Banking and Financial Services, April 23, 1998; and Insurance Services Office, Inc., *Financing Catastrophe Risk*, pp. 26-33.

ing activity above $10 billion in losses was minimal. (Another trading venue, the Bermuda Commodities Exchange, started in 1997 but shut down two years later because of a lack of activity.)

Catastrophe options differ from traditional reinsurance in their use of an index of losses for one of nine regions and states, rather than for a particular insurer, as a payout threshold. Using an index of industrywide losses significantly reduces moral hazard and adverse selection, but it creates basis risk for an insurer. Because the options contracts are not designed to match the losses of any individual portfolio, which may be more concentrated in one region than another, insurers will be exposed to the possibility of a mismatch between their losses and the contract's payout. Thus, insurers who purchase the options have an incentive to issue policies uniformly over a region, potentially lowering their cost of bearing risk.

Recent research indicates that such hedging of basis risk in catastrophe options for Florida insurers is effective only for the largest insurers and for smaller insurers that are highly diversified throughout the state. Creating intrastate index contracts (as opposed to statewide contracts) by separating Florida into regional zones would reduce basis risk and make the contracts more attractive to primary insurers.[14] However, regional subdivisions would raise transaction costs and reduce liquidity.

Catastrophe options are at a disadvantage compared with other financial instruments used by insurers in terms of their supply and demand. For insurers, the cost of the options cannot be deducted from income until the options are exercised or expire, whereas premiums paid for reinsurance may be deducted immediately. Furthermore, the National Association of Insurance Commissioners does not treat the options as reinsurance; consequently, insurers cannot increase the level of coverage they have issued after purchasing options. In addition, although the Chicago Board of Trade's clearinghouse guaranteed the options, some people doubted its ability to cover them following a major event.[15] For investors, the high cost of becoming informed about catastrophe risks may reduce their willingness to supply funds. (That drawback is present in most capital market products for disaster risks.)

## Contingent Notes

Debt financing helps insurers avoid financial distress following a catastrophe. Before an event, insurers agree to sell, and purchasers agree to buy, a debt issue (a note) at a fixed price. That arrangement for contingent financing allows insurers to issue debt at a specified rate following a disaster, when the insurer's financial condition might otherwise preclude such a sale. (If no disaster occurs, no debt is issued.) To guarantee the availability of funds, an insurer may require a potential bond purchaser or investor to hold funds equaling the amount of the note in government securities in a trust account; in the event of a catastrophe that met agreed-upon criteria, those funds would revert to the insurer. Investors receive a higher rate of return, or an up-front fee, to induce them to commit funds and to compensate them for the risk of interest rates rising or of only partial repayment. Since 1995, three of this type of transaction have been recorded, involving a total of $585 million in postdisaster bond placements.[16]

## Contingent Equity

A catastrophe equity put is a type of option that grants an insurer the right to sell a specified number of its equity, or stock, shares at a set price in the event of a catastrophe that exceeds a certain magnitude. Insurers would use the money they received from the sale of those shares to pay claims. Investors in the put option would receive a payment from the insurer to compensate for the risk that the price at which they agreed to buy the shares would be greater than the stock's market price. To the extent that the value of the stock after a disaster was less than the

14. See Cummins, Lalonde, and Phillips, *The Basis Risk of Catastrophic-Loss Index Securities.*

15. For an explanation of this default risk, see Jaffee and Russell, "Catastrophe Insurance, Capital Markets, and Uninsurable Risks," pp. 220-222.

16. Insurance Services Office, Inc., *Financing Catastrophe Risk,* pp. 23-26.

price at which the put option was exercised, the insurer would be partly hedged against catastrophic losses.[17]

For example, one insurance company paid investors about $2.4 million a year for the right to sell them $100 million in shares of preferred stock. The company could exercise that right if its losses from a single disaster exceeded $200 million or if its overall losses for any year exceeded $250 million. The dividends that would be paid on the preferred stock were tied to the London interbank offered rate, or LIBOR, plus a spread based on the company's credit rating at the time of the stock issue.[18] Because losses never reached the trigger level during the coverage period, the put options were never exercised.

One drawback specific to contingent equity is that investors are exposed to the general business risk of the primary insurer.[19] That is, a catastrophe might occur after a period in which the insurer's stock has performed poorly for reasons other than huge losses from a catastrophe. For example, a firm's management could change or its strategies could fail, making its stock price fall. Between 1996 and 1998, three catastrophe equity put deals were negotiated, with obligations to purchase stock totaling $250 million.[20]

## Swaps

Swaps allow insurers to trade exposures and diversify their holdings, thus reducing the risk that they could become insolvent after a disaster. Swaps do not change insurers' cash flows. For example, a Florida-based insurer might swap a portion of its hurricane exposure for some earthquake risk from a California-based insurer. Swaps may be negotiated over the counter or through the Catastrophe Risk Exchange, an electronic listing service started in October 1996 for insurers and reinsurers. The exchange will randomly select the policies to be swapped, preventing insurers from trading only their riskiest holdings.

Most large swaps are negotiated directly between companies that understand the risk of counterparty default. For example, Swiss Re and Tokio Marine & Fire Insurance, two of the largest global reinsurance companies, recently agreed to a $450 million swap. They engaged in three risk exchanges of $150 million to cover losses from natural perils: a Japanese earthquake for a California earthquake; a Japanese typhoon for a Florida hurricane; and a Japanese typhoon for a French storm.[21]

17. "Knock out" provisions, which eliminate the option if the postloss equity value is sufficiently low, can be embedded in the put option. See Doherty, "Financial Innovation for Financing and Hedging Catastrophe Risk."

18. Froot, *The Evolving Market for Catastrophic Event Risk*, pp. 8-9.

19. See Christopher M. Lewis and Peter O. Davis, "Capital Market Instruments for Financing Catastrophe Risk: New Directions?" *Journal of Insurance Regulation*, vol. 17, no. 2 (Winter 1998), pp. 110-133.

20. Insurance Services Office, Inc., *Financing Catastrophe Risk*, pp. 34-36.

21. Swiss Re, "Swiss Re and Tokio Marine Arrange Unique USD 450 Million Cat Risk Swap" (press release, Zurich, July 12, 2001), available at www.swissre.com (media center).

# Disaster Insurance Programs in Japan, Great Britain, and New Zealand

Most developed countries have government-backed programs that provide insurance against natural disasters. Similarly, most European countries have established, or plan to create, government reinsurance programs for terrorism risks.[1] Most of the programs provide limited coverage for property owners in the event of a disaster, generally at subsidized prices that make few adjustments for risks. The programs have two significant drawbacks: their subsidies discourage risk mitigation, and their setup does not allow the programs to receive the full benefits of sharing disaster risks internationally, through global reinsurers. That lack of risk diversification is especially evident in Japan's earthquake insurance program, Great Britain's terrorism reinsurance program, and, to a lesser extent, in New Zealand's earthquake insurance program.

## Japan's Earthquake Insurance

Although Japan has a government-backed insurance program (the Japan Earthquake Reinsurance Company), property owners in that country bear most of the risk of earthquakes. Despite the high risk of quakes, however, most homeowners choose to remain uninsured. In part, that choice may reflect the government's regulation of insurance rates, differences between risks and premiums, the taxation of capital reserves, and restrictions on market entry by foreign insurers, all of which limit the coverage that is available. Optimal risk spreading would dictate that most of Japan's earthquake risk be diversified globally. Such sharing of risks could ease the burden on the government's finances from a catastrophe in the $100 billion range, which could weaken the Japanese economy while leaving the government liable for most of the costs and claims.

In January 1995, an earthquake struck Kobe, Japan, destroying about 100,000 houses and buildings and resulting in 6,500 deaths and more than $110 billion in damages. Of that amount, insured losses were just over $6 billion.[2] (That magnitude-7.2 quake was considerably stronger than the quake that struck Northridge, California, in January 1994.) Less than 5 percent of residents had earthquake coverage, at least in part because the re-

1. See the statement of Thomas J. McCool, Managing Director, Financial Markets and Community Investment, General Accounting Office, before the Senate Committee on Banking, Housing, and Urban Affairs, published as General Accounting Office, *Terrorism Insurance: Alternative Programs for Protecting Insurance Consumers*, GAO-02-199T (October 24, 2001).

2. That loss was about 2.3 percent of Japan's output in 1995 but less than 1 percent of its stock of capital. For an analysis of the economic effects of the quake, see George Horwich, "Economic Lessons of the Kobe Earthquake," *Economic Development and Cultural Change*, vol. 48, no. 3 (2000), pp. 521-542.

gion was thought to be at low risk.[3] Losses were also large because of the number of old and substandard structures in Kobe. The government picked up many of the post-disaster costs, subsidizing about 90 percent of the cost of rebuilding public facilities.[4] To help the victims, the government also provided relief grants, low-cost loans, housing assistance to low- and middle-income residents, and tax breaks.

Since 1996, the Japanese government has reinsured the Japan Earthquake Reinsurance Company. The government bears about 85 percent of the liability for the company, which is capped at about $18 billion (1,800 billion yen). When the massive Kobe quake hit, the earthquake endorsement covered roughly 30 percent to 50 percent of a property's replacement value; claims were limited to less than $100,000 (10 million yen) for property and an additional $50,000 (5 million yen) for contents. The policy covered only minor additional living expenses. Premiums, which are generally more than twice as high as those in California, are based on the 500-year record of earthquakes in Japan and include a risk load. Only about 7 percent of Japanese homeowners choose to purchase earthquake insurance (16 percent in Tokyo.)[5] The government does not cover commercial insurance lines, but global reinsurers absorb some of that risk.

After the Kobe earthquake, the government increased the maximum amount of insurance that residents could buy. But the government still classifies risks incompletely. Rates for coverage distinguish between only two types of dwellings in four rate zones. Furthermore, they do not adjust for the age of a building, soil conditions, or the structural quality of surrounding buildings.[6]

## Great Britain's Terrorism Reinsurance

The British government created Pool Re as a reinsurer of last resort in 1993 after private reinsurers reduced their coverage for risks from terrorism following bombings by the Irish Republican Army (IRA).[7] With Pool Re, which is mutually owned by participating insurers, primary insurers may reinsure their risks from terrorism for commercial property losses and losses from business interruption. The program does not cover personal losses or workers' compensation claims. Pool Re sets premiums on the basis of the amount of insurance coverage, geographic location, and other risk factors; primary insurers set coverage limits. Pool Re must reinsure all offered polices. (That coverage begins after primary insurers pay roughly the first $150,000 in claims.) In turn, the British government reinsures Pool Re and agrees to guarantee any loans or lines of credit that Pool Re might seek. Once the pool's reserves exceed $1.5 billion, then it will pay to the government the greater of 10 percent of the net premiums remitted each year or a payment geared to the government's past losses. As of January 2001, Pool Re's reserves were nearly $1 billion.[8]

The government accepts liability for all of the claims above Pool Re's ability to pay. While the pool has had numerous claims—including one for about $150 million —there has yet to be a draw on the Treasury or premiums paid to the British government. (Pool Re would levy a special assessment on all participating firms before relying on the government's backstop.) Annual dividends are paid to participants each year that there is an underwrit-

3. More homeowners purchased coverage after the earthquake, a situation that is also typical in the United States. Coverage rose from 7.2 percent to 10.8 percent in Japan. Larger increases occurred in the areas hit hardest by the Kobe quake. See Tsuneo Katayama, "The Kobe Earthquake and Its Implication in Earthquake Insurance in Japan," in *Report on the January 17, 1995, Kobe Earthquake, Japan* (September 1997), available at http://incede.iis.u-tokyo.ac.jp/reports/Report_15/Katayama.pdf.

4. EQE International, *The January 17, 1995 Kobe Earthquake*, EQE Summary Report (April 1995), available at www.eqe.com/publications/kobe/economic.htm.

5. Ibid.

6. Katayama, "The Kobe Earthquake and Its Implication in Earthquake Insurance in Japan," p. 259.

7. For a more detailed description of Pool Re, see United Nations Conference on Trade and Development, *Comparative Examples of Existing Catastrophe Insurance Schemes*, September 29, 1995.

8. Tillinghast-Towers Perrin, "Pool Re and Terrorism Insurance in Great Britain," October 2001, available at www.towers.com/towers/services_products/Tillinghast/update_pool_re.pdf.

ing surplus—that is, premiums exceed claims. The dividend is 10 percent of the annual underwriting surplus.

Through Pool Re, terrorism insurance is widely available throughout Great Britain (although some commercial properties have decided to self-insure). Rates have dropped sharply as the IRA's terrorist activity has declined; no claims have occurred since mid-1996. Annual premiums have ranged from about $530 million in the early years of the program to less than $75 million in 2000.[9]

Critics have raised several concerns about Pool Re.[10] First, premiums are crudely set. There are only two geographic zones—the commercial districts of all major cities including London, and all other regions—and just a single adjustment for "target risk."[11] Rate schedules for the cities are three to five times those for the other zone. Target risk is related to a property's prominence and visibility. Buildings in London that are determined to be high risk are assessed an extra premium of 50 percent. Second, the premiums do not include any risk loads to compensate the British taxpayers for the risks that they accept. In fact, the British government explicitly subsidizes Pool Re through a 3 percent tax on all household and vehicle policies. (The larger subsidy is implicit.) Third, the government prevents risks from being diversified internationally by creating strong incentives for British insurers to reinsure only through Pool Re. This retains all terrorism risk within Great Britain, which may be an inefficiently diversified pool. Fourth, the program initially offered no incentives for risk mitigation. Companies that installed security cameras or hired extra guards, for example, were not able to share in the lowered risk through reduced premiums. That policy has been changed, and discounts are now available.

9. Ibid.

10. William B. Bice, "British Government Reinsurance and Acts of Terrorism: The Problems of Pool Re," Comment in *University of Pennsylvania Journal of International Business Law,* vol. 15 (Fall 1994), p. 441.

11. Northern Ireland is covered by a different program. The government directly reimburses terrorism losses and does not assess any premiums.

## New Zealand's Earthquake Insurance

The government of New Zealand insures residential properties against natural disasters through its Earthquake Commission. In addition to damage from earthquakes, policies cover damage from volcanoes, tsunamis, floods, and landslides. Although rates are not actuarially set and do not reflect risks, the program's fund balances have grown significantly in the absence of a major disaster.

The Earthquake Commission charges a flat rate of $0.05 for $100 of coverage, up to a maximum charge of $50, plus $10 for contents and a $7.50 tax.[12] (All amounts in this section are given in New Zealand dollars.)[13] The rate, which has not changed since 1944, was originally based on premiums charged by New Zealand's War Damage Commission. The government caps individual claims at $100,000 (and $20,000 for contents), roughly the replacement cost of a typical three-bedroom house in New Zealand. Private insurers provide coverage for exposures over $100,000 through their homeowners' policies.

No major earthquake or volcanic eruption has occurred in New Zealand since the government started insuring against disasters. The largest claim stemmed from a 1987 earthquake, which cost about $170 million (in current dollars). However, most of that cost was a result of commercial claims, and the government no longer offers commercial coverage. In most years, claims have been under $10 million. Only about half of the claims have been attributable to earthquakes; most of the remainder have resulted from landslides.

The earthquake fund has more than $4 billion in assets, primarily in the form of nontradable government securities (the fund can invest up to 35 percent of its balances in global equities). The Earthquake Commission has also purchased reinsurance of $1.5 billion in the private marketplace, improving its risk diversification. Estimates sug-

12. Personal communication to the Congressional Budget Office from David Rafferty of the New Zealand Treasury, March 18, 2002.

13. At exchange rates on August 21, 2002, a New Zealand dollar was worth about $0.47 in the United States.

gest that claims from a major earthquake hitting the country's capital of Wellington could range from $3.5 billion to $6.8 billion. If claims exceeded the commission's capacity to pay, New Zealand's treasury would provide additional financial support.